The Encyclopedia of

How It's BUILT

Thomas Telford designed the towers of Conway Bridge to match the nearby castle, built around 1284. These wrought iron chains are his originals, installed in 1826.

The Encyclopedia of

How It's BUILT

Edited by Donald Clarke

A&W Publishers, Inc.

New York

*Most American houses are built of frame-timber. This
historic example, in Cambridge, Massachusetts, was the
headquarters of General Washington from 1775 to 1776 and
later the home of the poet Longfellow. It shows the typical
horizontal weatherboard style.*

First published in the United States of America in
1979 by A & W Publishers, Inc.
95 Madison Avenue
New York, New York 10016
By arrangement with Marshall Cavendish Limited

This book may not be sold outside of the United
States of America and its territories, Canada and
the Philippine Republic.

Library of Congress Catalog Card Number:
79–51558

ISBN: 0–89479–047–1

Printed in Great Britain

Introduction

Everything we see in our industrial society is designed and built by other people before we get a chance to use it. Cups, buildings, meadows, even food are preceded to some extent by models, drawings or otherwise clear intentions.

For example, the so-called 'wilderness areas' of the United States exist by design, and quite precise maps of the way they must be kept are on file. The reason for this is partly that a few unspoiled acres will promote greater efficiency by refreshing the work force—or, since very few people actually go there, by providing refreshing images. Even politicians, to the extent that they are social or economic planners, may be considered designers and builders—hence phrases like 'architect of foreign policy'.

All this designing and building is a bit frightening, when we consider that other people are making decisions about what kinds of tools, houses, ornaments and even recreation we have available to us. But there are two ways in which we can salvage our self-respect as consumers: if we want to, we can repair our own cars, paint our own walls and bake our own bread. And we participate in the design process everytime we buy something, for when we choose a set of dishes or a pair of trousers, we are *not* choosing all the others. In our market orientated economy, consumers still have a way of developing a wariness that will put a designer out of work or a builder out of business if he doesn't do his job well enough.

This series of books, *How It Works, How It's Made* and now *How It's Built*, has been dedicated to the proposition that we will get more satisfaction and value for money to the extent that we are aware of the forces behind our choices as consumers. *How It's Built* will tell you why your neighbourhood looks the way it does, and even gives you a history of the designing and building which led to the world in which we live today.

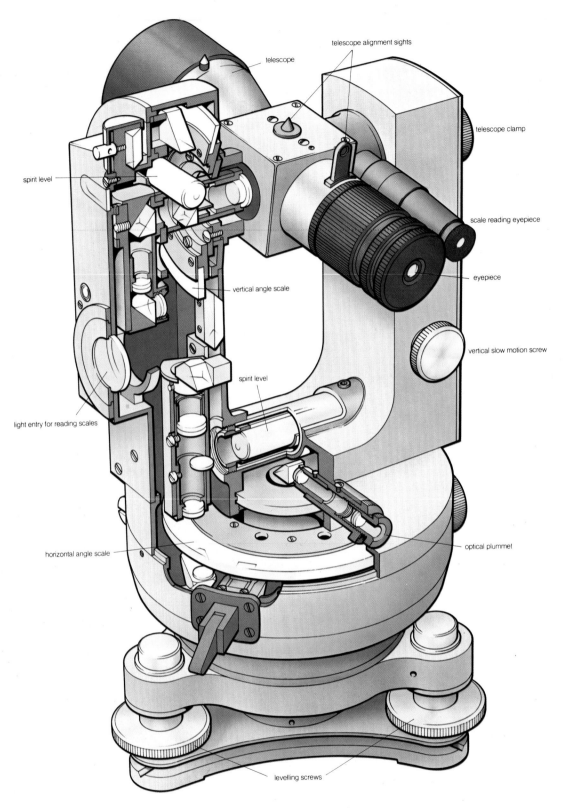

telescope alignment sights

telescope

telescope clamp

scale reading eyepiece

spirit level

eyepiece

vertical angle scale

vertical slow motion screw

light entry for reading scales

spirit level

horizontal angle scale

optical plummet

levelling screws

A typical surveyor's theodolite. The scales are read through an eyepiece positioned next to the main telescope eyepiece. The vertical and horizontal angle scales, which are made of glass, are illuminated by daylight directed through the window on the left by means of a mirror (not shown). The light is split into two beams which are conducted to the scales and then into the telescope head by a system of lenses and prisms.

Contents

The Arch 9
Assembly lines 12
Boatbuilding 22
Bridges 28
Building, history 39
Building, modern 44
Cathedral construction 49
Construction machinery 55
Construction materials 74
Demolition 87
Design 90
Drainage 102
Flood control 105
Frame construction 109
Land reclamation 113
Lighthouses 116
Mining and quarrying 120
Pyramids and stone circles 130
Roadbuilding and street lighting 139
Rubbish disposal 148
Sewage and water supplies 151
Ships and docks 158
Surveying 166
Transportation of buildings 168
Tunnels 170
The Vault 176
Wells 180

The Arch

The simplest way of spanning an opening in a wall is by means of the *post* and *lintel*, in which a single solid member (the lintel) is laid horizontally with its ends resting on vertical supports (the posts).

If for some reason lintels are unobtainable then a way must be found of spanning openings with an assembly of small pieces somehow jointed together firmly enough to stay permanently in place. The result is an arch, which has the further advantage of being able to span greater distances than a lintel. Its history can be traced back to places such as the Tigris and Euphrates valley where timber was scarce and large flat stones were expensive.

An arch, therefore, is an opening spanned by a collection of wedge-shaped pieces, the *voussoirs*, which stay in position by pressing in on one another. The joints between them appear to radiate from some central point lying within the opening, and sometimes from points which lie outside, so every type of arch has a characteristic curvature. The simplest and visually most natural shape is the semicircle but other designs have been used.

The central voussoir at the arch's apex is the *keystone* or headstone, traditionally the last to be set into position to 'lock' the whole thing into a strong and stable structure. A keystone is not always necessary: there may be a join at the apex instead, as is common in Gothic arches. Gravity tries to pull the headstone downwards, but the thrust is carried on either side by the voussoirs immediately flanking it. These in turn have their total thrust carried through the whole semicircle of pieces in a sideways direction until it reaches the vertical part of the wall and can descend directly to the foundation. In short, an arch works because vertical weight is deflected into sideways thrust and transferred to the walls.

Because of the sideways thrust the arch is not a stable structure in itself: that thrust tries to make the bottom of the structure spread out on either side. To stop this happening there must be enough solid material at the sides to act as flanking buttresses.

For this reason an arch is more naturally placed within the body of a wall rather than at either end. Series of arches are suitable for bridge building or aqueducts because the river banks or valley sides make excellent buttresses. Similarly, long colonnades consisting of repeated arches, favoured by the Romans, need sizeable lengths of unperforated wall at each end to bear the combined thrust of the entire series, though intervening posts or piers can themselves be quite slender.

This tenth century gateway from the Mayan city of Chichen Itza, Mexico, used the corbelled or false arch, so called because it transmits thrust straight down instead of spreading it.

It is interesting that even when builders took care of thrust by incorporating tie-bars in the structure so that large, solid flanking piers became unnecessary, their omission proved visually unsatisfactory. It was as if the eye unconsciously visualizes the thrust exerted by an arch and expects to see it contained in something solid.

Solid flanks are unnecessary where colonnades are completely circular, for their entire weight becomes a single, unified downward thrust. The most remarkable example is the Colosseum at Rome, consisting of three storeys of circular arch colonnades surmounted by a visually solid fourth-storey wall (added later). Faith in the stability of the arch is well expressed in the line 'When falls the Colosseum, Rome shall fall'. Built between 70 and 82 AD, it is still standing.

Arches were first developed by the Assyrians and Babylonians, and reached high levels of technical complexity under the Romans, who seem to have borrowed the idea from the Etruscans, who inhabited Italy before them. (The ancient Greeks, however, bothered little with arches, concentrating instead on transforming the crude post and lintel concept into superb colonnades of columns and pediments.) This phase of arch history saw efforts concentrated on two problems: deciding which type of arch would be most stable in each case—always rather a rule of thumb affair owing to the complexity of stress calculations—and simplifying the temporary wooden framework needed for support during construction.

Although arches were developed highly in the Gothic and Renaissance periods of architecture, particularly in the construction of vaults, in the 20th century they have almost been forgotten. This is because modern building techniques such as prefabrication and the use of squared-up structural frames infilled with non-load-bearing walls, coupled with the expense of craftsmen, have made arches uneconomic.

Opposite page: the ruins of the huge barrel vault at Ctesithon, now Taq-e-Kisra, in Iraq. The brickwork follows a natural parabolic curve like that of a modern concrete arch or vault. These ruins have now been completely washed away by floods.

Right: the famous but useless Gateway Arch at St. Louis, Missouri, a parabolic arch made of reinforced concrete.

Below: several types of arch. The post-and-lintel as well as the corbelled are not true arches. The centres from which the arches are drawn are shown as red dots, and the spring line is also red.

Nowadays, however, people are reacting against the appearance of the 'giant matchbox' buildings, and there has been a realization that curved shapes are both attractive and feasible using materials such as concrete, plastics and laminated wood. They can also be structurally more stable than rectangular styles.

Modern analyses of thrusts and stresses in arches and their supports show that they tend to follow a parabolic curve. Hence in theory the structurally most natural form for an arch is a parabolic curve going from foundation to foundation without intervening piers. Such shapes are easily achieved in reinforced concrete and laminated wood—indeed, they are more efficient and economic forms for these materials than those of the squared-up approach. Furthermore, with laminated wood the problems of centering practically disappear, since the structure is light enough to be built flat on the ground and then hoisted up into position. Arches of this sort are used to span extremely wide spaces.

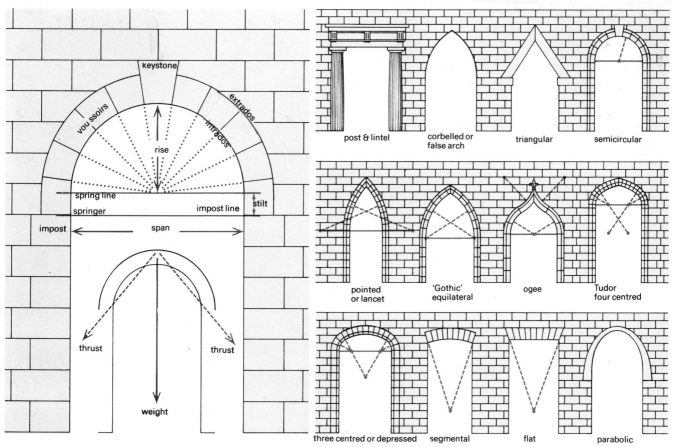

keystone

voussoirs

extrados

intrados

rise

spring line

springer

impost line

stilt

impost

span

thrust

thrust

weight

post & lintel

corbelled or false arch

triangular

semicircular

pointed or lancet

'Gothic' equilateral

ogee

Tudor four centred

three centred or depressed

segmental

flat

parabolic

11

Assembly lines

One of the earliest examples of mass production was in 16th century Venice, where a reserve fleet of galleys was kept ready in prefabricated sub-sections, ready for assembly as needed. But true mass production could not begin until the 19th century. British engineers such as Henry Maudsley had perfected precision machinery, especially screw-cutting lathes, so that parts could be made which were interchangeable. One of the first uses of modern methods was when Samuel Colt and Eli Whitney (who also invented the cotton gin) used interchangeable parts to make guns.

The assembly line is the basis of modern mass production methods, enabling large quantities of goods to be produced at reasonable cost. Everything from transistor radios to typewriters and television sets to warships has been produced on assembly lines; indeed, it was the

Above: an assembly line in the British Aircraft Corporation's factory producing front fuselage sections for the BAC/Aerospatiale Concorde supersonic airliner. There are final assembly points in both England and France, using parts and sub-assemblies from more than 800 sub-contractors. The airframe is assembled using overhead cranes, the enormous wiring harness is installed, then the plane is mounted on its undercarriage and the engines are fitted. Next it is painted in the colours of the customer airline and moved outside, where the instruments and cabin furnishings are installed.
Right: a simpler assembly line for electric typewriters. An operator takes a machine from the roller conveyer, adds a component and pushes it to the next station.
Between these two extremes, the possible variation of the assembly line method is limitless, from simple conveyers to complete automation.

assembly line technique applied to the production of ships, aircraft and other weapons which boosted American production capacity beyond anyone's expectations during World War II.

The best-known example of assembly-line production is undoubtedly the car. The techniques first perfected by Henry Ford have brought car ownership within the reach of nearly everyone in industrialized countries, and have profoundly changed the nature of the society in which we live, for better or worse.

The basis of the assembly line is repetitive work. Each worker does the same simple task over and over again as the work moves past on some sort of conveyer. Time-study analysis is important here; the various jobs must be as efficient and economical as possible, and many things must be taken into consideration. For example, it may seem obvious that the worker who installs the headlamp on one side of the car has enough time to install the other one as well, but the time it takes to cross in front of each car and back again has to be taken into consideration.

In one case, a time study expert decided that both headlamps could be installed by one person. He neglected to take into consideration that at that point on the final assembly line there was a pit underneath the car where other people were working. The employee involved would have had to jump back and forth across the pit all day.

Time study, or work study, was the brainchild of F. W. Taylor, an American who was a friend and associate of Henry Ford. Like many idealists, his faith in human nature was touching, his knowledge of it somewhat inadequate. He treated the human body as though it possessed no consciousness, assuming that the average worker was not very bright, yet expected the worker to share in the increased profits accruing from more efficient work. The dehumanizing effect of repetitive work is only now being dealt with by means of experiments in work sharing, team work and so forth. And the natural tendency of many early industrialists was to keep all the profits for themselves. One of the most important functions of the labour union is the organization of the workers so that a balance of power can be maintained, yet for many years such unions were not even legal. In any case, there can be no doubt that Taylor and Ford between them invented a

Diemakers finishing a 25-ton die for drawing floor pans for cars. Final grinding and polishing are done after some test pieces have been drawn. The machines which press the sheet metal parts are as big as houses.

technique which has resulted in an outpouring of consumer goods and a material standard of living unique in history.

Sub-assembly

A car factory contains a large number of assembly points. The large parts of the car, such as the body and the engine, are built on assembly lines of various sizes and delivered as sub-assemblies to the final assembly line.

In the engine division of the factory, for example, there is an automated machining line called the in-line transfer system. A typical installation for machining cylinder blocks for Austin cars is 65 ft (20 m) long and has fourteen stations. Its operation commences with a casting picked up by an air-operated carriage which places it on a transfer bar conveyer. The piece is located at the first station. The tools advance, carry out their machining operation, and retire. The piece is then advanced to the next station, located, machined and so on until it reaches the end of the line. Meanwhile other castings are following it. A finished block comes off the end of the line about every four minutes. Such an installation may perform drilling, reaming, tapping, boring, honing and milling operations with only two operators, one to put the casting on the line and one to take off the finished block.

In the meantime, other parts are also being prepared. The pistons are turned in automatic lathes, and are graded according to their precise diameters. On the engine assembly line itself, the blocks are usually bolted to a fixture which can be rotated for the convenience and efficiency of the assemblers. The completed engine, built up from the machined block with pistons, connecting rods, crankshaft, valve gear and various external fitments such as water and fuel pumps, is then ready to go to the final assembly line to be installed in a car.

The body line

In the body division, the large sheet metal parts of the car, such as the roof, door panels, floor pan and so forth, are delivered by conveyers from the press room. These

13

conveyers usually consist of hooks on chains which are pulled along the ceiling by electric motors. There are miles of such conveyers in a car factory, and they carry everything from baskets of parts such as pistons to the car itself on the final assembly line.

At the beginning of the body line, sub-assemblies of the floor and the sides of the car are made by placing the sheet metal parts in *framing bucks*, where they are held in place by clamps while they are spot-welded together. Spot-welding is a resistance welding process. The welding instrument looks like a large G-clamp with copper tips; it is water-cooled inside. When the operator presses the button, the clamp closes and a surge of electrically produced heat flows through the two layers of sheet metal, welding them together. A finished car will have thousands of spot-welds in it. Sparks fly in every direction; safety goggles are absolutely necessary. People who do such work learn not to carry matches in their shirt pockets.

The sheet-metal sub-assemblies are then installed in a larger framing device with clamps sticking out all over it. These framing bucks are then pulled by a conveyer down the first of the assembly lines, called the body line. The roof and the rear wings [called fenders in the USA] are all spot-welded onto the car, along with various framing parts. The front fenders are often screwed on for ease of replacement in case of damage. Sometimes the sub-assemblies of the doors, the bonnet [hood] and boot-lid [trunk-lid] are also added at this point; or sometimes they are painted separately and added later.

The car as described is of *unit body* construction. Nowadays most cars are built this way, but a few cars (and most cars made in former times) have separate frames, made of heavier steel, to which the completed car body is bolted, using rubber grommets to prevent squeaking in the finished product.

When the body is complete it goes into a tank full of special paint which rust-proofs it. The paint is sometimes applied electrostatically; that is, the tank contains paint and water, and the steel is electrically charged so that the particles of paint are attracted to it.

When the rust-proofing primer is dry, the car is spray-painted in the specified colour. The various coats of paint are often dried by passing the body through a drying oven, but a recent development is radiation curing, in which a specially formulated paint is 'set' almost instantly by infra-red or ultra-violet radiation.

Opposite page: fitting a spot-welded 'uniside' into a framing buck with the floor pan at American Motors.
Below left: finishing Volkswagon roofs in a factory in Brazil. Tiny dents and dimples must be removed.
Below centre: floor pans clamped in jigs.
Below right: spot welding on the body line.

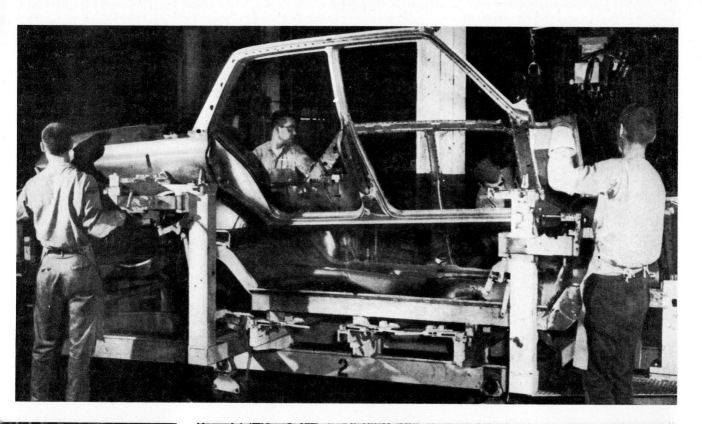

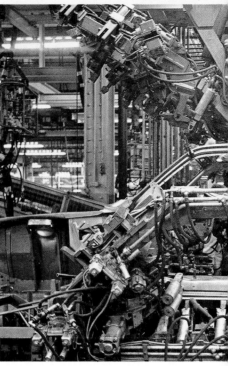

The trim can be added as soon as the paint is dry. Each carmaker has its own way of building the car; the trim line may be a small assembly line between the body line and final assembly, or all the trim can be added at the beginning of the final assembly process. In either case, the 'trim' can be loosely defined as those parts of the car which are not part of the drive train. It includes any chrome strips or name plates fastened to the outside of the body, which are usually attached by means of clips through holes in the sheet metal. It includes all of the interior upholstery: carpets or rubber mats on the floor, seats, padding on the instrument panel, door panels, headliner and so forth. Window glass, door handles and their associated mechanical linkage, the fuel tank and many other parts may be installed on the trim line. The steel part of the instrument panel may be part of the body, with padding, instruments and so forth added later, or the entire instrument panel may be pre-assembled and added on the final assembly line.

Final assembly

The final assembly line is the culmination of the whole process. Here all of the sub-assemblies come together, nowadays often directed by computer, and the cars drive away from the end of the line under their own power. A single assembly line may produce 1500 cars a week, taking anywhere up to 90 minutes travel on the line for each car, depending on how much trim remains to be added at this stage. If the final assembly line includes virtually all of the trim, it can be nearly a mile long. Parts and small sub-assemblies from all over the factory are delivered by conveyer to the final assembly department, carefully timed to arrive at the right time, or they may be delivered in steel tubs or on wooden or steel pallets by fork lift trucks. Many parts of the car are not made by the car factory at all, but are bought from outside suppliers, which may or may not be partially owned by the car manufacturer. These items vary from one manufacturer to another, but can include the instruments, steering column, lights, radiators, gearboxes, and the *harness*, which is the wiring assembly and is usually mounted on

the firewall behind the engine. Part of the assembly process is the connecting of the wires to terminals for lights and instruments all over the car.

The underside of the car is fitted out by the workers in a pit underneath the car, including the exhaust system, tubing for fuel and brake fluid and so on.

The most dramatic part of the final assembly line is where the engine joins the rest of the car. Again, each manufacturer has his own way of doing it. One typically convenient way of doing it is as follows: the rear axle assembly complete with differential, the drive shaft and the engine arrive separately at a small sub-assembly area next to the main line, where they are bolted together. (The gearbox is already bolted on to the back end of the engine.) Then they are dragged around a curving section of a conveyer, on rollers, which steeply rises as the assembly mates with the underside of the car. Assembly-line workers are standing ready to fit nuts, washers and other parts on to downward protruding bolts as the heavy steel plate on to which the engine is mounted fits over them.

In one car factory some years ago, a group of employees tried to obtain a car loaded with every conceivable optional extra at a stripped-down price, and with an employee discount at that. The computer-printed broadcast sheet attached to the bumper of the car gave no hint of what was going on; the plan required the cooperation of a great many people up and down the line. It was only foiled because the 'souped-up' engine wouldn't fit into the car at that point in the assembly, causing a great deal of confusion. In the end, several people lost their jobs.

If the assembly line has to stop for any reason during working hours, a light goes on in the offices of the higher-ups, who immediately want to know what the problem is. For 'down time' results in thousands of people standing around doing nothing, and getting paid for it. Long before any assembly begins, an order commissioning the car has been printed by the computer, and the colour, trim specifications and other information about the finished car are already known. Cars for export may have

Above left: a framing buck at British Ford's Dagenham factory. The parts are all held in place by clamps while spot-welding takes place.
Above centre: a 'multiwelder' at Ford's Halewood plant. Banks of spot-welding electrodes close around the body.
Above right: not all the welding is spot-welding. The console in this American Chevrolet Camaro is welded into place with an electric arc.

Below: a completely automated body assembly line at a Fiat plant in Turin, Italy. There are eighteen of the robot welders, which can do 500 welds without human help, except for maintenance. Automation in many industries is only just getting under way, and will pose social and economic problems which must be solved. The wizards at Fiat have built one of the largest car factories in the world under contract in the USSR.

Opposite page, top to bottom: Pontiac bodies being inspected between baking operations at a Fisher body plant in Michigan; a car body being immersed in water containing an anti-rust paint, after phosphate treatment to clean it. The metal is charged so that the particles of paint are attracted and evenly distributed all over it; and spraying a British Ford with acrylic enamel paint.

Left centre: spot-welding British Fords at Dagenham.

This page, below: engine assembly in a Leyland factory.

Bottom: Oldsmobile V-8 engine blocks coming from the transfer bar, still wet with coolant. There may be more than a hundred machining units; the hydraulic tubing and cylinders which operate the machinery can be seen on both sides.

left-hand or right-hand drive; they may have specially-built engines for countries where high-octane fuel is not available. The combinations of specifications to which a car can be built are almost without limit. These are some of the reasons why the arrival at the assembly point of the right parts at the right time is one of the most important functions in the car factory.

All up and down the line, power tools, usually operated by compressed air, are used to tighten up nuts and bolts. Each tool is adjustable to provide the right amount of torque. If the tool is a large and heavy one, it is suspended from the ceiling on a pulley, and counter-balanced (this is true of the spot-welding device in the body division as well).

As the car reaches the end of the assembly line, the wheels are put on and the nuts tightened with a power tool; the radiator is filled with year-round coolant and sealed; and some fuel is pumped into the tank. (This last operation is carried out in a strictly no-smoking area.) Then, if all goes well, the car starts.

The car is given a check-out on a dynamometer, which in this case consists of rollers built into the floor so that the car can be 'driven' without going anywhere. The brakes and brakelights are tested here, as well as the gearchanges.

The headlamps are adjusted, all the lights are checked, the steering alignment is set up, and the engine fitments are adjusted on the 'tune-up' line. If there is anything seriously wrong, the car goes to a final repair department; otherwise it is driven to a company parking lot to await shipment to a dealer. Transport charges within the country of origin are usually the same within a wide area, so that customers who happen to live far away from the factory are not unduly penalized. Some car companies will allow a customer to pick up his own car, thus avoiding transport charges altogether.

Above: crankshafts are handled by transfer equipment.
Below: British Vauxhall cylinder heads being machined in a transfer bar.
Opposite page, top: the motor pit on the final assembly line at American Motors' Kenosha, Wisconsin plant.
Opposite page, bottom: a new Cadillac, fresh off the line, being test-driven on a dynamometer.

Boatbuilding

Boats developed from a number of primitive water craft in different places and at different times. The earliest such craft may have been a log used as a float to cross a river. The log became a dug-out canoe when fire and axes were used to hollow it out. Two such canoes were lashed together for stability, or an outrigger was added. In pre-historic Europe, stability seems to have been achieved by placing weights in the bottom of the canoe.

In Egypt, where timber was not available, boats were built by lashing together bundles of reeds, and pulling the two ends of the boat up from the water by means of ropes. In China, flat bamboo rafts were in use for river haulage, probably from about 4000 BC. Later, perhaps about 1000 BC, they began to adapt these rafts

by laying the bamboo along the curved sides of semicircular wooden planks, thus creating a vessel which had a number of solid bulkheads down its length. This continued to be developed until it resulted in today's Chinese 'junk'. Another ancient type of boat, the wooden frame covered with skin or bark, survives today in the form of the Irish curragh, the Eskimo kayak and the American Indian canoe.

For thousands of years, wood was the most common material used in boatbuilding, but in the 1950s glass

Below: boatbuilding is one of Man's oldest crafts. On Lake Titicaca, in Peru, the traditional boat of the local fisherman is made of reeds lashed together, and is remarkably similar to ancient Egyptian boats.

reinforced plastic brought rapid changes, allowing mass production techniques and greater flexibility in hull design. Today about 70% of boats built in Europe and the USA are made of GRP.

Plank construction

Boatbuilding in wood is a skilled craft involving the use of a great many wooden components to build a watertight structure. This has to combine stability in the water with the ability to withstand stresses often comparable with those experienced by jet aircraft.

Wood boatbuilding follows two principal styles, clincher (sometimes spelled clinker) with overlapping planking, and carvel or caravel where the planking is smooth. The clincher or lapstrake technique gives a 'monocoque' or stressed skin construction. The shape of the craft is developed by the addition of successively fastened planks, sometimes using only a single mould or guidance shape amidships. After the planking is

completed, light frames are steamed or sawn to shape and added to the interior to strap the planking together as a safeguard against a plank splitting along the grain. This construction makes the boat very light but vulnerable to damage. The overlapping edges or lands of the planking are liable to wear and it is difficult to keep the planking watertight once it has been disturbed. It is therefore normally used only for small craft such as beach boats.

In carvel construction wooden planks are fitted edge to edge over a completed framework which determines the shape and forms the structural support of the finished craft. The framework consists of a centreline piece called the keel with a companion piece called a keelson, a stem at the front which is joined to the keel with a wooden knee (angle piece) called a foregripe. At the other end the upright part of the framing consists of a sternpost with a supporting knee. The framing across

Right: the two classic ways of building small wooden boats are the carvel and the clinker method.
Next page: Uphams shipyard, Brixham, England in 1956. The Mayflower II was timbered in the traditional way.

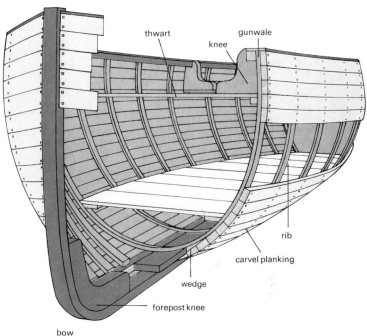

bow

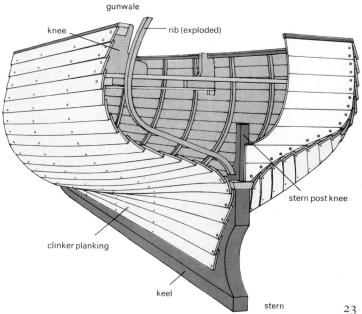

the boat is normally built with timber sawn from branches whose grain lies roughly in the required curves. These frames are in turn strapped together inside with full length wooden planks called stringers, and at the deck edge with an inwale or interior plank which is sometimes called a beam shelf if it has to carry the deck beams. Originally a number of heavy planks called wales were arranged longitudinally and fastened to the outside of the frame before the skin planking was fitted. Their function was to stop the planking from spreading when the caulking compound was hammered between the planks to make the hull watertight. With modern improvements in building techniques, wales have largely disappeared.

There are many variants of the two techniques and light, steam-bent strap frames are often used to augment and lighten the sawn frame structure. Other common variations include composite constructions where steel frames are used with a wooden keel and planking; multiple planking, where two or more layers are placed diagonally to make a strong skin which is more watertight, though difficult to repair, and strip planking. Very narrow planks are used in the normal planking but are nail fastened through their thickness to the previous plank as well as being fastened to the frames.

To get the best performance from a fast power boat a knuckle or chine (projecting corner) must be built at the division between the bottom and sides of the hull. This chine is built like the centreline keel and the technique is known as a chine construction.

Plywood

During the 1920s boatbuilding techniques using plywood were developed, largely in the USA. The invention of waterproof resin adhesives led to the development of water resistant 'marine grade' plywood. This quickly became popular for building bulkheads or vertical wall divisions of the hull and later it became a very common material for the planking skins of inexpensive small craft. This technique, still widely used in 'do it yourself' built craft, was the first real attempt to use glue to stop water from entering between the components of the hull.

Despite great ingenuity, the shapes which can be constructed from bending flat sheets of plywood are very limited and were invariably angular in appearance. The next step therefore was to make the actual plywood sheets over a curved mould of the required shape. This type of building is now used for high-quality racing yachts where the high costs involved can be balanced against the strength and light weight. This moulded plywood construction involves the planking of the mould with very thin wood sheets of veneer thickness. These are added in successive glued diagonal layers until a finished thickness approximately half that of a carvel planked hull is achieved. If cold setting adhesives are used each layer is held down with staples until the glue sets. For hot moulded construction the layers are held in place by air bags or a vacuum press and moved into an oven to give a quick 'cure' to the hot setting glue.

A variation of this construction is the use of laminated structural members. Here the keel, stem, beam, and ribs are built up over a mould to the required dimensions and curves.

The moulded plywood hull of a racing yacht after curing in an oven.

Glass reinforced plastics

The most popular form of construction today uses polyester resin reinforced with glass fibre, generally spoken of as fibreglass or glass reinforced plastics (GRP). The normal building process starts with the construction of a full size solid model of the final boat, called the plug. Over this a hollow mould is formed by a similar process to the hull construction. The hull and deck mouldings, and even mouldings for the interior accommodation, are formed in separate moulds.

The materials used in GRP construction consist basically of polyester resin reinforced with finely spun glass fibres either in cut pieces or made up into a woven cloth.

The mould is first coated with wax or some other release agent to prevent the resin from adhering to the surface. Next a gel or surface coat of resin, usually impregnated with pigment to suit the final colour scheme, is sprayed or painted onto the surface of the mould. A very light supporting mat of glass fibres is then placed on top of the resin and pressed in with a roller until completely saturated with wet resin. Subsequent layers of moulding resin with heavier glass reinforcement carefully rolled into it are added until the required hull thickness and strength are attained. The resin hardens in three stages: first to a soft gel, then to a point where the moulding can be removed from the mould, and then over a period of weeks it matures to full strength. Bulkheads and reinforcements are added to the moulding either in or just out of the mould. Considerable skill and care are necessary in glass fibre construction in order to make certain that the resin is properly supported by glass at all corners and edges, that no air is trapped between the layers and that the glass is thoroughly saturated with resin.

Another common method of fibreglass construction uses a hand spray gun to deposit both the resin and chopped glass strands which leaves a mat of partly saturated fibre that must then be rolled to complete saturation as before. This method involves skilled spraying and accurate control of quantities to ensure uniformity. Glass fibre construction is also mechanized in other ways, such as the use of glass mats previously saturated with resin, and vacuum or pressure resin saturation of previously laid glass reinforcements.

The undoubted advantage of the glass reinforced plastics method of construction is the monocoque or stressed skin nature of the hull, without any joins where water might enter. Another benefit is the reduction of maintenance, which can be one tenth of that for a

Above, top: a ship's hull of fibreglass, or glass reinforced plastic. This material has gained enormously in popularity since World War II.

Above: 'plating' an aluminium racing yacht.

Left: glass fibre matting being pressed into the wet resin.

conventional wooden hull. Further benefits lie in being able to use the moulds for production building, and the ease of achieving a high finish. The material, however, is fairly heavy (flotation chambers are often included) and inconveniently flexible. One way of correcting these faults is a 'sandwich' construction where another material is placed between the GRP layers to improve stiffness and reduce weight. End grain balsa wood slabs or foam plastics are commonly used over areas or in patterns as required during the course of the hull moulding.

The cost of building a single hull in glass fibre is very high owing to the cost of the plug and mould. To overcome this some craft are built with a PVC or polyurethane foam core planked over moulds like a traditional wooden hull and then covered with fibreglass inside and out. The outside surface of the hull is ground, sanded smooth and paint finished.

Other plastics boatbuilding methods include simple foam plastic castings and vacuum forming, where a plastic sheet material such as ABS or polyethylene is heated until soft and then sucked down over a hull-shaped former. This is either thick enough and the right shape to be rigid in its own right, or it may be formed of two or more mouldings which are filled with injected foam to give buoyancy.

Other materials

Boats are also built in metal. Steel, for instance, has been a common material for building boats as well as ships. Its strength and weight characteristics limit its use to larger craft but with welded construction and the new anti-rust coating, steel has become more versatile.

Aluminium is also a popular material for high performance one-off yachts, increasingly since new alloys have reduced the original serious corrosion problems of aluminium in salt water. The metal's lightness also makes it suitable for small craft which have to be manhandled. Some small aluminium hulls are made by stretch forming, where a sheet of material is stretched bodily into shape over a hull-shaped former.

Above: a concrete boat sounds like something that would sink like a stone, but in fact they are cheap and durable. A hull made of concrete can be cast in a form or it can be made practically by hand; see also picture on page 81. Concrete is one of the most versatile of building materials.

Below: modern lightweight materials have not taken all the hard work out of sport sailing: that would spoil the fun!

Bridges

Probably the earliest type of bridge was the beam bridge where a stone or tree was laid across a stream. These would have been forerunners of the stone 'clapper' bridges in and around Dartmoor, England, or the timber 'clam' bridges of south east Cornwall. The first suspension bridges were festoons of creepers, cane, bamboo or vines, tied to tree trunks and hung across a gorge, similar to those still to be seen in China, India and Africa at the present time. There are three types of bridge, *beam, arch* and *suspension*, and the vital differences among them are that beam or girder bridges simply rest on their supports; arch bridges are in compression and thrust outwards on their bearings at each end; and the main element of suspension bridges, the cables, are in tension and exert a pull on their end anchorages. These three types of bridge may be combined to assist each other in composite structures, of which there are a great many examples.

Beam or girder bridges may be joined together over the piers, when they become continuous; or they may be further modified to form cantilever bridges. There remain pontoon bridges and movable spans, both of which are essentially beam bridges. An early description of a pontoon bridge refers to one built across the Hellespont by the Persian king Xerxes, using fifty-oared ships and triremes (a type of large warship) as floats. The life of pontoons, however, is short and their maintenance a constant problem, so that they are unsuitable for permanent structures. Swing spans, vertical lift bridges and bascules (a type of drawbridge raised and lowered with a counterweight) did not come into use until the 19th century.

The finest early bridges were undoubtedly the circular masonry arches of the Romans, some of which, like the 98 ft (30 m) spans of the Alcantara bridge over the Tagus, have stood for nearly 2000 years. Two important contributions by the Romans were, first the discovery of a natural hydraulic cement called *pozzolana* (so named because it came from Pozzuoli near Naples), which

enabled them to make lime mortar or concrete for use in underwater foundations; the other was the development of *coffer-dams*, made by driving timber piles to form an enclosure around the proposed site of a timber pier in midstream, then draining out the water inside, so that the pier could be built in the dry. River piers presented such a problem to early bridge builders that the Shouster bridge, built in the third century AD over the Karun River, Persia, is not straight, but winds across on 41 piers that were built on rocky outcrops in the river.

From the fall of the Roman Empire until the Renaissance, fortified bridges were built, often with chapels, shops or toll houses on them. Pointed masonry arches were generally used on mediaeval bridges, as on Old London Bridge, which was completed in 1209 and survived for more than 600 years. Some bridges, however, such as the Pont d'Avignon which crossed the Rhône in 1187, on some 20 spans of 100 ft (30 m), had lofty elliptical arches. Bridge building gradually became a science following the construction of the famous masonry bridges of the Renaissance, such as the Rialto bridge in Venice and the Santa Trinita bridge in Florence, and aided by the work of scientists such as Galileo and Robert Hooke on the theory of beams and structures. The 18th century saw bridge trusses of timber, first evolved by Palladio, developing into the American covered bridges, and witnessed the world's first all-iron bridge, the semicircular arch of 100 ft (30 m) span built at Coalbrookdale, England, in 1779.

Iron and steel

Bridges consisting of numerous cast iron arches of increasing span began to appear in the 19th century: the first major iron truss bridge with pin connections was built in the United States and the first iron cantilever bridge in Germany. In 1850 the Britannia railway bridge, which was the prototype of the modern box girder, consisting of two wrought iron tubes through which the trains ran, was opened across the Menai Straits. Nine years later this was followed by Brunel's famous Royal Albert bridge at Saltash over the Tamar estuary in Devon. In 1879 one of the great bridge disasters occurred, when 13 wrought iron spans of the new Tay railway bridge were blown down, no special provisions at that time having been made to assess or resist wind pressures. Telford's Menai suspension bridge (1820–1826) of 580 ft (177 m) span, with chains of wrought iron links, survived for 115 years. Wire cables instead of chains were first used in France, but many early suspension bridges collapsed in storms or under repeated rhythmic loads.

The mass production of mild steel plates and sections for shipbuilding led to their use in bridges, the first big steel crossing being the Eads bridge with three arches, each over 500 ft (152 m) long, which was completed in 1874 over the Mississippi at St Louis, Missouri. In 1883 the Brooklyn bridge of 1595 ft (486 m) span, designed by

Opposite page: a double-arched Roman bridge in Switzerland, one of many monuments to Roman skill.

Left top: a clapper bridge in Dartmoor, England, probably built by the Celts. 'Clapper' comes from the mediaeval Latin 'claperius': a pile of stones.

Left centre: a rope and log bridge in Mexico.

Left: a primitive cantilever bridge in Afghanistan.

29

John A. Roebling, was successfully completed in New York. The four main cables are each built up of parallel wires of galvanized, high tensile steel 0.19 inch (4.8 mm) thick with an ultimate strength of 71.5 tons per sq inch (11,000 bar). They were spun in place by the method used in principle in all the big suspension bridges built subsequently in the USA, and steel stiffening trusses were incorporated in the deck to damp out oscillations caused by wind or traffic. Before the end of the century, mild steel's ductility, workability and ultimate strength of 30 tons per sq inch (4600 bar) in both tension and compression, led to its being widely used throughout the world. The Forth railway bridge (1881–1890), with two cantilever spans of 1710 ft (521 m) each, was the first long span railway bridge to be built of steel, and was designed in the shadow of the Tay bridge disaster to withstand wind pressures of 0.39 psi (27 mbar). Today high tensile, structural steel suitable for electric arc welding and fabrication by flame cutting is widely used, and friction grip bolts have superseded riveting and site welding.

Reinforced concrete

The earliest bridges of reinforced concrete date from the end of the 19th century and the principal spans were all arches. Among the most notable are the Sandö bridge (1943) in Sweden with a span of 866 ft (264 m) and the Gladesville bridge (1964) in Sydney, New South Wales which has a span of 1000 ft (304 m). After World War II the process of prestressing concrete, that is putting it into a state of compression by tensioning steel bars or wires that passed through it, led to marked economies in bridge design. In this period improvements in concrete technology led to its strength being doubled so that working stresses up to 3000 psi (207 bar) can now safely be adopted. By saving one third the volume of concrete and three quarters of the weight of steel reinforcement that would otherwise be used, prestressed concrete bridges are now competitive with other materials and widely adopted for spans up to 600 ft (183 m) or more. Notable recent cantilever·spans are the new London Bridge (1973), the Eastern Scheldt bridge (1965) in the Netherlands and the Bendorf bridge (1964), West Germany, which has a span of 682 ft (208 m). By employing cable bracing, however, prestressed concrete bridges can be considerably increased in span, the longest to date being the Wadi Kuf bridge (1970) in Cyrenaica, Libya, which has a span of 983 ft (300 m).

Bridge foundations

Pneumatic *caissons*, large open bottomed cylinders sunk onto an area dredged level and pumped out with compressed air, allow foundations to be built on a river bed. They were first used in Britain on the Rochester bridge in 1854. They were subsequently widely adopted for depths up to 120 ft (37 m) and used on many great bridges, such as Brooklyn and the Forth. Men cannot safely work in compressed air at depths greater than 120 ft (37 m), however, so for deeper foundations a tube is first sunk and then cleared to form a well shaft using an open grab machine, before depositing the concrete. The trend now is to abandon pneumatic caissons and support bridge piers on numbers of long, large diameter piles or thin-shelled cylinders. The piles may be 3 ft (1 m) or more in diameter and 200 ft (51 m) long. In the USSR thin-shelled prestressed concrete cylinders have been used from 5 to 20 ft (1.5 to 6 m) in diameter, with walls 3.5 inch (89 mm) thick, and sunk by means of vibro piledrivers to depths of over 100 ft (30 m).

Notable 20th century bridges

Post-war reconstruction and the building of modern roads have given rise to the greatest bridge building activity in the history of the world. In 1918 the Quebec bridge, a steel cantilever of 1800 ft (549 m) span, was completed after two tragic failures. The great Sydney Harbour bridge was opened in 1932 and has a steel arch of 1650 ft (503 m) span. This was erected over deep water as two cantilevers until they met and joined in the middle. The United States led the way in the construction of many great suspension bridges, the longest of which exceed 4000 ft (1220 m) in span. The most recent of these, the Verrazano Narrows bridge (1964) at the entrance to New York Harbour, has steel towers 680 ft (207 m) high and carries twelve lanes of traffic on a span of 4260 ft (1298 m). Owing to the curvature of the Earth's surface, the tops of the vertical towers are 1.6 inch (4 cm) farther apart than their bases—and the length of steel wire used to form the four main cables would stretch more than halfway to the Moon.

Beam bridge

The most common form of bridge construction is the *beam*. A plank placed across a ditch is a simple beam bridge, and the action of the load is resisted by bending stresses in the material. For a *simple supported* beam, which rests on two supports only, any load acting downwards puts the bottom part of the beam in tension and the top part in compression. These stresses develop sufficient leverage (the technical term is 'bending moment') to support the load. The forces keeping the beam straight must, by a fundamental law of statics, equal the load tending to fold it up. Therefore for a given amount of material, a deep beam will give more leverage than a thin one, and material near the top and bottom faces works to better effect than material in the middle. Beams are usually made in an I- or T-shaped cross-section.

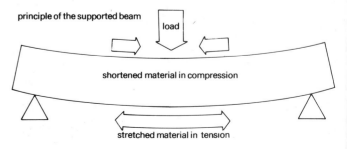

principle of the supported beam

load

shortened material in compression

stretched material in tension

Opposite page, top: the Garabit Viaduct over the Truyere, designed by Alexander Gustav Eiffel.

Above: a reinforced concrete arch in West Germany.

Below: Sydney Harbour bridge, designed by Sir Ralph Freeman and completed in 1932.

Cantilever and suspended span

In a simple supported beam the maximum bending moment occurs at the centre, diminishing to zero at the ends. This is inconvenient if a slender, graceful appearance is wanted. The *cantilever and suspended span* arrangement has the cantilevers and suspended spans separated by 'hinges', which divides the bending moment into two parts, *sagging* as before in the mid-span regions and *hogging* (tension at the top, compression at the bottom) over the inner supports. If desired, the hogging may be arranged to take more than half the total moment, which fits well with a graceful curved profile—Waterloo Bridge in London is an example. The Avonmouth Bridge near Bristol in western England is similar, but is a *continuous girder* bridge with no hinges. This is now common practice, despite the fact that if one pier sinks slightly this bends the beam and adds to the stresses.

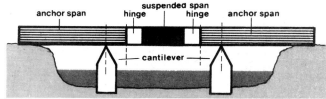

Suspension bridge

The suspension bridge can cover a vast span but has a serious drawback: it is very flexible and traffic loading may produce a large deflection, particularly when it acts near the quarter point of the span. For this reason a stiffening beam or box girder is almost always provided to supplement the cables. Even so, this type is rarely used for railway bridges as trains are heavier, and their loading more concentrated, than road traffic. Suspension bridges are, however, the solution for the longest spans, since wire for the cables can be produced cheaply with much greater strength than the steel suitable for beam construction. A compromise which is now popular for intermediate spans is the *cable-stayed* bridge which has stays joining the towers and deck.

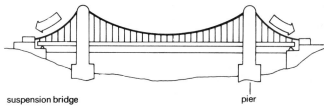

The arch

In bridge construction the arch could be considered to be a suspension bridge turned upside down. Like the suspension bridge the arch puts a much larger force on the abutments (the end supports) than a simple beam does, although it pushes rather than pulls. This makes the ground do part of the work, thereby saving material in the structure, but it requires reliable foundations. Flexibility is less of a problem in this type of construction because ordinary steel can be built up into a large arched rib or girder to resist buckling.

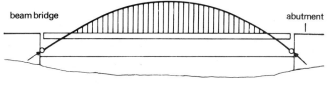

Box girder bridge

A box girder is a hollow girder of either rectangular or trapezoidal cross-section made with thin walls, designed to provide strength with minimum weight. It is usually made from a combination of mild and high tensile steel plates with stiffeners welded to their inner surfaces. Box girders are used extensively in modern bridge construction, giving slender lines and an economical use of materials. They are used for both road and rail bridges, sometimes with steel boxes in combination with a concrete deck. Cable-stayed bridges also have box girder decks. In modern suspension bridges such as the Severn in England and the Bosporus in Turkey, completed in 1973, box girders are used for the deck.

The first box girders were used by Robert Stephenson in 1849–1850 when he built the Britannia Bridge across the Menai Straits in Wales. This bridge had the trains actually running inside rectangular iron tubes. These tubular boxes acted as continuous beams with two spans each 460 ft (140 m) long, making the Britannia Bridge the largest beam bridge in the world. This bridge has since had to be rebuilt because of damage caused when the accumulation of paint, tar and cinders in the boxes caught fire. A smaller version at Conway in Wales, that Stephenson built to establish the technique, still carries the main line to Holyhead.

The only suitable material available to Stephenson was wrought iron, and this only in 18 inch × $\frac{1}{2}$ inch (46 cm × 1.3 cm) strips. So a very large number of strips had to be riveted together to make up the necessary strength.

Up to the mid-nineteenth century, steel was an expensive metal that could only be produced in small quantities, but the Bessemer process, invented in 1856, made mass production feasible. Cast iron is easy to pour into shapes, but is very brittle, difficult to produce in uniform quality, and unreliable in tension. Wrought iron is purer and more malleable (easily shaped) than cast iron but is very ex-pensive. Steel has the advantage of uniform quality, giving accurately calculable properties. Heavy steel elements of complicated cross section could only be economically mass produced if they were straight and with a standard cross section. This made it cheaper to join these in a skeleton forming an open truss framework rather than a large tubular box. Only since the mid-1940s has the great advance in electric arc welding made it feasible to join relatively thin plates.

Once a satisfactory method of joining plates had been developed the box offered several advantages. A truss bridge needs a separate deck with its own system of beams to take the loading, say from the wheels of passing vehicles, back to a suitable joint on the truss. But with a box, one piece of steel can do treble duty. It can function as part of the main beam, and can help to distribute the moving load, and it can be part of the actual deck surface. Another advantage of a box is great stiffness against torsion (twisting). Stresses arising through torsion are usually caused by lopsided loads, and a road bridge with one traffic lane blocked creates a considerable twisting action, as does a railway bridge with a train on one line. It is usually possible to design a box with a streamlined aerofoil cross-section so that the force exerted on it by the wind is very much less than the sum of the forces on the many parts of a truss doing the same job.

The problem with boxes is that thin plates tend to buckle when subjected to compression, perhaps at only a small fraction of the load that the same cross sectional area of metal could carry if constructed in a less slender form.

Stress

The slenderness of a box can be seen if a sheet of paper 8 inch (20 cm) wide and 0.004 inch (0.1 mm) thick is folded into a square box with 2 inch (5 cm) sides. This is equivalent to a box girder 10 m deep with *webs* (side walls) 2 cm thick. A box with plain flat plates would have a low resistance to buckling, so to improve this stiffeners are welded on, which increase the strength in the same

way as corrugated cardboard. Calculating the strength of a stiffened plate is extremely complex.

The strength of a plate is affected by the welding process itself, as the joining seam of metal is melted into place and then shrinks as it cools. Once the weld hardens, further shrinkage leads to stresses as the weld pulls against the surrounding structure, and this often causes significant distortion as well as permanent stressing. This problem is further complicated by the fact that structural steel is not simply elastic, but if overstressed can 'yield', so that plastic deformation (permanent bending without breaking) takes place while the material still succeeds in supporting the load. Limited yielding is often acceptable and can even out local peak stresses (stress concentrations) caused by welding or otherwise. It would not be economic to keep the peak calculated stress (including welding effects and every possible combination of loads) below the value at which any yield occurs. But a method of calculating the true strength, allowing for yield, still needs much research.

Box girders are easier to manufacture than built-up girders, and as they can be batch-produced, unit costs are kept down. The interiors of the boxes are not exposed, so they are easily protected, which helps to reduce maintenance. Boxes are often pre-assembled to produce large sections of deck that can be lifted in one piece. Where temporary support is not possible or where the bridge is too high, the deck can be erected by using the box girders already in place to form cantilevers. Each new section can be run out over the existing structure, and then placed in position on a temporary beam under the girder, which is moved along as construction proceeds.

In the early 1970s box girder bridges were the subject of a great deal of research and discussion, largely as a result of the collapse of four big bridges while they were still under construction. One of these bridges was in Britain at Milford Haven and another was the Lower Yarra Bridge, Melbourne, Australia. The others were in Vienna over the Danube and a bridge spanning the Rhine in Germany.

Most box girder bridges are much less spectacular

than these projects, as the box girders merely replace the I-section girders supporting a concrete deck for simple road or railway overbridges. A concrete deck is economic

33

for spans up to 500 ft (150 m), but for very long spans the higher cost of a steel deck is balanced by saving the cost of the supporting girders, as a steel deck is only one third the weight of a concrete one.

Box girder bridges are used mostly in Britain, Europe and Australia, and seldom in the United States. The Avonmouth Bridge in England, to carry road traffic over the Avon river below Bristol, has a steel deck (with a normal asphalt running surface) 132 ft (40 m) wide supported on two boxes 19 ft 6 in. (6 m) wide which are 25 ft (7.6 m) deep at mid-span. The main span is 570 ft (174 m). For even longer spans it pays to augment the girder with some form of supporting cable but the girder itself can still conveniently be made in box form. These cables do not contribute much to the twisting stiffness of the bridge, and as one big box is much stiffer against twisting than two smaller boxes sufficient for the same bending stiffness, a single box is more likely to be used. The single deck box girder for the Erskine Bridge over the Clyde in Scotland is 57 ft (17.4 m) wide (the deck overhangs this by 18 ft each side) and 10 ft (3 m) deep, with stay-cables to reinforce the 1000 ft (305 m) span. The Humber Bridge in northern England, which was scheduled to open in 1977, will be the biggest of them all. It is a suspension bridge with a clear span exceeding 4500 ft (1370 m) and a deck formed from a streamlined box girder 72 ft (22 m) wide by 15 ft (4.5 m) deep.

Right: a box girder bridge under construction in Brazil.
Far right: the Nanking bridge is a truss structure several miles long.
Below: a typical triple-cell steel box girder bridge.
Below right: statics is very important in the design of bridges and other structures; it enables the designer to calculate loads and stresses. The interconnecting triangles in this structure make it very stable.

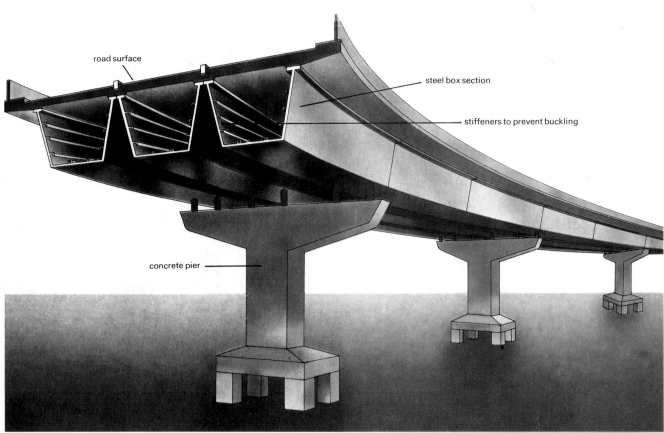

road surface

steel box section

stiffeners to prevent buckling

concrete pier

35

Notable bridges

From the viewpoint of the structural engineer it is the span, the distance between supports, that counts when designing a bridge. It is relatively easy to build bridges miles long by mass production of short spans. New high speed railways, such as the Japanese Shin-kansen system, are including hundreds of miles of such construction, with little of the interest and problems of the following examples.

Robert Stephenson built the Britannia Bridge over the Menai Straits in Wales in 1849–1850. This was a four span continuous beam bridge carrying a double track railway, each track running inside its own box-section girder. These boxes were made on shore by riveting together small strips of wrought iron, and were then floated into position to be jacked up to the top of the piers that had been prepared to receive them. The main spans were 460 ft (140 m) long.

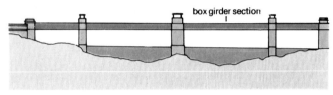

Brittania bridge – continuous beam

The Forth Railway Bridge was built by Sir Benjamin Baker and Sir John Fowler in 1883–1889. It was the first application of steel on such a large scale, and it remains today the most spectacular cantilever and suspended span bridge in Britain. The main spans are each 1700 ft (520 m) long. It was constructed by building up the trusses piece by piece on the site from quite small elements; more than 50,000 tons of steel were put together at some times employing 4000 men.

Forth railway bridge – cantilever and suspended span

Steel arches were popular for long spans in the early years of this century. The Hell Gate at New York was spanned in 1916 by an arch of 973 ft (298 m) that carried four railway tracks and had a very heavy loading. A similar design was chosen for the Sydney Harbour Bridge. This was built in 1925–1931, has a span of 1650 ft (500 m), and also carries a very heavy load on a deck which is only 137 ft (42 m) wide and includes both a main road and a suburban railway line. Even before it was completed, however, a comparatively lightly loaded road bridge was opened at Bayonne, near New York, with a span of 1652 ft (504 m).

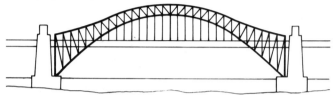

Sydney Harbour bridge —steel arch

The first bridge to span 1000 ft (305 m) was built over the Ohio River in 1849, and was also remarkable as a pioneering example of steel wire cables for bridges. The Ohio Bridge had the conventional suspension layout but for spans this long the modern solution is usually the cable-stayed bridge in which one or more sets of straight stays join the deck to the tower.

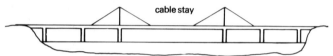

Erskine bridge – cable-stayed

The Avonmouth Bridge, opened in the 1970s, is a classic example of the elegant slender appearance that can be obtained by carefully varying the section of a continuous beam bridge. The deck is well over 100 ft (30.5 m) wide and is made from steel plate which works as part of the flanges (top surfaces) of the two boxes beneath. The main span is 570 ft (174 m) long.

Avonmouth bridge – continuous beam bridge

After the 1000 ft Ohio Bridge was built, and despite its collapse in a storm after only five years, the suspension bridge was used for even greater spans, largely due to the construction techniques developed in America, notably by Roebling. Perhaps the greatest landmark was the 4200 ft (1280 m) clear span of the Golden Gate Bridge, San Francisco, built 1933–1937. In the last twenty years British engineers have developed suspension bridges with a carefully streamlined box girder as the stiffening beam. The Humber Bridge for example, is designed to span 4625 ft (1410 m) with a clear opening of over 4500 ft (1372 m).

Golden Gate suspension bridge – San Francisco

Materials

The choice of material is basically between steel or concrete, although the distinction is not absolute, as all concrete bridges include a large amount of steel as reinforcement, and the majority of steel bridges have concrete decks.

Concrete is the cheapest serviceable material for the job; it can readily be made to have a *crushing* strength of 30 newtons per square millimetre, N/mm^2 (say 4500 pounds per square inch). As the actual strength on site varies rather a lot according to the care and supervision of mixing and placing the concrete, the designer would probably design for a stress of about 10 N/mm^2 under the nominal working load. Steel is substantially more expensive, but is made under carefully controlled conditions, so that the designer could work to perhaps 200 N/mm^2. Even so, an equivalent member might cost eight times as much in steel as in concrete. Other materials are even more expensive, and can rarely compete effectively with concrete and steel.

The great disadvantage of concrete is that its *tensile* strength (resistance to pulling) is very small. This does not matter much for arch construction, but for most other cases steel reinforcement must be added to carry all tensile forces wherever they may occur. Round steel bars are usually the cheapest and most convenient form of steel reinforcement. An alternative way to use reinforcement is to stretch it by hydraulic jacks before the concrete is poured around it—or more commonly for bridges, before it is *grouted* (glued with a cement paste) into place in *ducts* (long tubular holes) through concrete which has already hardened. When the jacks are released the concrete is thus *prestressed* by an equivalent compression, and under any subsequent loads behaves as if it were able to resist tension until the net combination of prestress

Opposite page: the Severn bridge from England to Wales, with a central span of 3240 feet (987m). The suspended deck is an aerodynamically-shaped steel box structure.

Right: a lifting bridge over the Harlem River, New York.

and stress due to the added loads exceeds the tensile strength. The principle is the same that one uses instinctively when lifting a whole row of books off the shelf in one unit by pressing them together. To be economic, prestressing requires higher strength materials than 'ordinary' reinforced concrete, but fortunately steel can be produced in suitable wires and small diameter bars to much greater strength than in other forms at only moderate extra cost.

Concrete is about four times heavier than steel per unit of strength. For short span bridges the traffic load is large compared with the structure weight, so this does not much matter. For long spans the converse is true; for 660 ft (200 m) span the structure weight would be perhaps three times any added load even with all-steel construction. Sufficient strength must be provided to resist the total effect so for such spans steel is also certain to be cheaper in total cost. Steel also gains because less material has to be handled, and supported during the construction process, and avoids or reduces the amount of *formwork* (mould and temporary support) required.

Erection

Much ingenuity is used in erecting big bridges. A good designer considers the characteristics of the site very carefully with erection in mind at an early stage of the design: can large pieces be brought into position from below, or from above, are very large cranes likely to be available, what is the premium for speed of erection?

The simplest method, perhaps, is to assemble a complete span in a workshop or on some convenient level ground and lift it into position in one piece. This is clearly most often suitable for small bridges, but the main span of the Ohnoura Bay Bridge in Japan, 640 ft (195 m) long and a mass of nearly 1000 tons, was recently pre-assembled in Tokyo, shipped complete several hundred kilometres, and landed gently on its piers as the tide went down.

The next easiest way is erection on *falsework* or scaffolding, where the height is not too great and there is no objection to the obstruction caused. Falsework can

become complex and very expensive, using tubular scaffolding or prefabricated metal panels like a scaled-up version of children's constructional toys. The most spectacular recent example in Britain was the Wentbridge Viaduct in Yorkshire, with a main span 310 ft (94 m) long and 98 ft (30 m) above ground; 125 miles (200 km) length of scaffold tube was used to make a complete working platform across the valley.

Finally the bridge can be built out piece by piece, cantilevering forward from the section already completed. For steel girder bridges the sections are usually pre-assembled into lengths of 49 ft (15 m) to 65 ft (20 m) and then carried forward and suspended on a carriage beyond the end previously reached. When the new unit has been bolted or welded into position the process can be repeated. This is a very tricky method technically, as the stresses induced are high and in many places quite different from those that will act in the permanent condition, so that there is every temptation to work to a narrow margin of safety to minimize costs. The structure may be alarmingly flexible when at full stretch just before meeting in the middle. The deflection of the two halves of the Wye Bridge, England—main span 770 ft (235 m)—due to their own weight was over 10 ft (3 m), which had of course to be allowed for to enable the parts to be joined smoothly.

Concrete erection
Similar methods can be used for concrete bridges although the sections must be much shorter—usually less than 16 ft (5 m) because of the greater weight of the material. Precast sections can be joined on by the 'pre-stressing' method, or alternatively the concrete can be cast on the spot in formwork (a mould) which is carried forward by a carriage similar to that used for precast or prefabricated steel units. The Medway Bridge in England (main span 152 m—500 ft) was built by the latter method, and was cantilevered in both directions away from the piers at the ends of the main span. To give stability, temporary (falsework) piers were placed in the side spans, but these could only carry quite small loads so on each pier the two working faces, one on each side, had to be kept precisely in step to preserve the delicate balance of up to 5000 tons of concrete.

Bailey bridge
Early in World War II it became apparent that existing military bridges were becoming obsolete because the latest armoured vehicles were too heavy to use them. The types of bridge used by the British army, for example, had a maximum span of 130 ft (40 m) and could not carry a load of more than 26 tons. Then in 1941, Sir Donald Bailey introduced a new design, which was easy to construct, adaptable, and capable of carrying up to 100 tons over spans of 220 ft (67 m). It was first used by the British in the North Africa campaign in 1942, and by the US army in Europe from 1944; during the war American and British firms made 200 miles (322 km) of fixed and 40 miles (64 km) of floating Bailey bridge equipment.

The basic Bailey bridge is made up of sections, each 10 ft (3 m) long, joined together to form a continuous span. Each section, known as a 'bay', has two side members (trusses) of girder construction, joined by cross pieces (transoms), which support the sheet metal roadway or 'decking'. To increase the load capacity of the basic bridge, extra trusses and sometimes transoms are added to strengthen it. The bridge is easily transported since it is brought to the site in component form, and can be assembled quickly and easily with a minimum of labour.

For crossing rivers the Bailey bridge can be supported on large plywood floats known as 'pontoons'. These are composed of three sections, each 20 ft (6 m) long, which are joined end to end to make one 60 ft (18 m) pontoon. The number of pontoons needed to support the bridge depends on the width of the river and the load to be carried. The floating part of the bridge is not one long rigid span, but is made up of 32 ft (10 m) lengths, each supported by a pontoon at either end with an additional one at the centre for heavily loaded bridges. These sections are joined together by connectors which allow a certain amount of flexibility between them, to prevent excessive stress in the trusses as the pontoons rise and fall when a vehicle passes over them.

The end spans which connect the floating section to each bank are called truss hinge spans, and as their name implies they act rather like hinges in order to allow for any variation in the height of the river. The pontoons at the ends of the floating section are braced by a distributing girder because of the extra strain placed upon them by the hinge spans.

As well as being useful during war, the Bailey bridge has proved invaluable in peacetime, particularly for replacing bridges destroyed by accidents or swept away by floods. Development work still continues, but the basic design has not radically altered from the original. An aluminium version was produced in the USA but it did not gain widespread acceptance, and steel is still the standard material used. Most of the improvements that have been made involve the methods of assembly, with more emphasis being placed on the use of machinery instead of manpower.

British troops scrambling across a Bailey bridge in Germany in April 1945.

Building, history

Man's basic needs were originally food, shelter and clothing, but as the population increased he was forced to move away from his caves and consider alternative shelter to protect him from the heat of the sun, the rain and the cold. Screens and huts of branches and twigs, walls of turf, and slabs of stone were used, depending on the natural resources available.

New techniques depend on the development of man's intellectual ability to isolate the technical problems and then to evolve the skills which are necessary to solve them. The development of suitable tools for extracting and handling building materials is equally important.

The first recognized surge forward in building technology occurred about 3000 BC. Before then, man's technical abilities were restricted by the simple and relatively inefficient tools. (Constructional materials such as the sun-dried hand-moulded mud bricks used in the Near East from about 6000 BC did not require the use of tools.)

By about 3000 BC, men had discovered bronze. This alloy, capable of taking a sharp and lasting edge, was ideal for the production of saws, axes and chisels. Such tools enabled trees to be easily felled, trimmed and squared for structural use. Stone, previously used in its natural form, could now be cut from the natural rock bed, squared, and tooled smooth ready for polishing. The timber houses of this period which evolved in Northern Europe and the vast stone structures of the Nile Valley resulted from this simple but important metallurgical discovery.

Two further discoveries which occurred between 3500 and 3000 BC were responsible for furthering technological progress. The first of these was the development of the wheel. The second discovery was the art of communication through word-signs, which first appeared in Mesopotamia about 3500 BC making it possible for new ideas and techniques to be disseminated. The art of kiln-fired pottery now spread to brickmaking and durable clay bricks made in a mould became available in Mesopotamia from around 2500 BC. The huge *ziggurats* (places of worship) built about 2000 BC were of brick set in bitumen.

Greek building

About 320 BC the Greeks made important contributions to technology, largely through the Museum, essentially an institute for teaching and study, which was founded by Ptolemy at Alexandria. They evolved among other inventions systems of compound pulleys for lifting heavy weights, as well as the application of geometry and trigonometry to linear and volumetric problems. The Romans seizing on the former devised vast machines for lifting weights and for driving piles into the earth to support foundations for heavy buildings and civil engineering structures. Geometric science, freely available to Arabic scholars in the original Greek, was not available to Western scholars, architects and craftsmen until translated from Arabic into Latin early in the 12th century. This opened the way for more exact and scientific mathematical calculations.

A reconstruction of a British Iron Age roundhouse at a museum in Worcestershire. It may have been a smokehouse.

By 300 BC the Greek city states had at their disposal building techniques which enabled their artisans to cut, transport, carve, polish and erect marble buildings of remarkable beauty and precision. The structures were themselves simple in conception incorporating simple post and lintel techniques, relying on gravity and metal cramps for stability. Columns were built up of drums which were held in position with pins fitted into sockets at the centre. Some of the columns used for the Parthenon were 6 ft (2 m) across. The roofs of these great buildings were low pitched with terracotta or marble tiles supported by timber framework.

Roman and Byzantine

While the Greeks built in solid stone and marble, the Romans used these materials solely as a facing or permanent shuttering to their buildings, forming the core of the structural wall or vault of concrete. Concrete was available to the Romans because of the strong hydraulic cement which was known as *pozzolana*. Because their buildings were larger and the spans of openings and vaults greater, flat stone lintels could no longer be used to support the loads imposed upon them. Now semicircular relieving arches, vaults and domes were used to enclose space and support heavy wall structures. Fired in kilns similar to those employed in the 19th century, Roman bricks were used extensively both as a facing material for the concrete core and also as a backing for the decorative mosaics or painted plaster finishes. Small multi-coloured mosaic tiles were used widely during the 5th and 6th centuries AD to decorate the interiors of Christian Byzantine churches. These buildings, constructed mainly of brick, sometimes with columns or piers of stone employed a novel building technique whereby the domes, which were such a feature of the interiors, were constructed by *corbelling* out each successive course until the dome was complete.

The principal problem which concerned Byzantine builders was the construction of the spherical dome on top of a square base. In early small structures this problem was overcome by means of a squinch, a small semicircular vault placed in the corner of the square base which carried the springing of the dome. With larger structures this was not sufficient and the *pendentive*, a series of spherical triangles supported and springing from a square base plan and carrying the ring of masonry which supported the dome itself, was evolved. Most buildings employing this technique are to be found in the Western Mediterranean and Slav countries, but the technique was revived again in various churches built around the 12th century in Aquitania in Western France.

Western European developments

The fall of the Roman Empire caused a serious setback to the progress of building technology. Fortunately isolated pockets of culture remained where the old skills survived, in Germany under the Ottoman Empire and in France. The art of brickmaking and the construction of brick buildings survived in central Europe slowly spreading west until it found its home in the Low Countries. There an abundance of clay suitable for brickmaking and a dearth of stone led to the growth of a fine tradition of brick building. In France a relatively stable government under Charlemagne kept alive the stone mason's art.

The advent, in France, of new cultures from the North gave impetus, after a period of time, to the old traditional techniques. The Northmen who settled in Northern France adopted the heavy structural walls and round arch and adapted these for their own use. Their adoption of Christianity enabled this to flower from the beginning

of the 11th century into a great series of Romanesque cathedrals and churches, which spread across Western Europe meeting a similar cultural movement from Germany and crossing the Channel into England after the Norman Conquest. Most of these buildings had timber roofs but it was not long before all were roofed with, at first, barrel vaults and later groined vaults. The Roman walling technique was continued with a facing of dressed stone. The core of good concrete was, however, replaced by stone waste from the dressing floor filled in with weak lime mortar resulting in walls with little structural strength.

Without strong Roman cement to help them, Romanesque builders found the enclosure of space with wide span vaults difficult. The geometry of the round arch was also a hindrance, restricting the plan of the vaulted bay to a square. The problem was solved in the construction of the new cathedral at Durham in England, begun in 1093 AD. The high vault of the nave incorporated for the first time the pointed arch for the transverse ribs and the wall arches enclosing the clerestory. This new building technique was soon copied at Winchester, at Peter-

Above: a 14th century Florentine ceramic plaque depicts stonemasons at work. The scaffolding consisted of wooden poles lashed together.

Right: Durham Cathedral, built in the 12th century, is a fine example of the skill of Romanesque builders.

borough and at Selby. Similar experiments were being conducted in France where rib vaults appeared in the Abbey of St Etienne at Caen. The first use in England of the Gothic sexpartite vault, which developed from these works, was in the new choir at Canterbury constructed towards the end of the 12th century. The new Gothic building techniques were based on an increasing effort to lighten and simplify the masonry structure and to replace the solidity of walling by coloured glass. The removal of structural solidity required its replacement by buttresses, pinnacles and piers to accommodate the lines of thrust from the high vaults and transmit the loads safely to the ground.

Up to the 15th century most domestic buildings were constructed with a timber frame. None have survived, however, so this knowledge is derived from manuscripts and archaeological investigation. Infilled squared timber framework was the usual constructional technique employed in Roman country villas and farmhouses. The *cruck* house developed during the eighth and ninth centuries AD, and was in use up to the 15th century. The lack of planning flexibility caused the cruck to be replaced by the regular timber frame which developed into the balloon frame in the 16th century as is used still in modern timber construction. Some domestic work was constructed of stone following the ecclesiastical constructional and stylistic fashion of the period. In urban areas however, huddles of squalid timber framed houses persisted well into the 18th century.

In Italy, Gothic had never been happily absorbed. The

Italians were always conscious of their imperial past. The writing of Plutarch in the 14th century and the rediscovery of the writings of classical antiquity led to the emergence of the Renaissance in Italy. Probably the first building to be designed in the form of the classical renaissance was San Lorenzo in Florence about 1425. The new technique employed round-headed windows, barrel and groined vaults, and semicircular domes, all borrowed from Roman sources.

The style spread across Europe, and coinciding with the Reformation in England, was manifest in the great Elizabethan and Jacobean houses of the period. Inigo Jones and Christopher Wren were the greatest exponents of the new classicism which continued throughout the 18th century. New materials were, however, used to simulate the old. At St Paul's, Wren used timber supported on a brick cone to provide the outline of the dome. Painted stucco replaced the coursed ashlar masonry of earlier days. Structurally solid walls and timber floors continued to support loads in a traditional fashion. Only in America, with their vast resources of timber, were new structural forms employed for these classical structures, surviving today in the elegant timber-framed weatherboarded houses and mansions of New England and the East coast of the USA.

Far Eastern techniques

Building in the Far East evolved its own style but in China very little exists which is older than the 12th century: the Great Wall of China, completed in about 210 BC, was restored in the 16th century.

Hard rammed earth bases sometimes faced with brick formed the base of Chinese buildings. Timber columns in pine or cedar set in stone bases were placed above this and tied together by a system of beams. To support the roof, beams placed one above the other, and of diminishing size, were used—unlike the triangulated truss system developed in the West. Brackets (*tou-kung*) were used to extend the rafters (often of bamboo) beyond the columns. Chinese buildings are remarkable for their variety of roof designs. There were three basic roof types: gabled, hipped, and pyramidal. Roofs were clad in tiles sometimes glazed in beautiful colours—blue, yellow or green.

Japanese building techniques were influenced by the building materials available such as timber, clay, fibres and metal, and to a certain extent by the natural hazards of their environment such as earthquakes and storms. Designs such as the pagoda were imported from China.

Indian building which began in the Indus valley about 3000 BC to 2000 BC was founded on large brick cities laid out in the grid-pattern and built of brick. This was followed by a simpler period around 100 BC when buildings consisted of timber posts and beams pegged together with bamboo and enclosed with wattle and plaster or even brick walls. Stone, often beautifully carved and massed in pyramidal structures, was used for temples between the fifth and eighth centuries.

The legacy of the past

In modern times it has become traditional for architecture students to travel all over the world looking at buildings in every style: for the first time in history the design of buildings has become consciously eclectic, and all of the styles we have seen in this chapter appear in bits and pieces everywhere. Now that we are entering an era of so-called post-modern architecture, it remains to be seen whether architects will be able to incorporate in their work the integrity of the builders of the past.

Above: Chinese architecture did not change dramatically from the 14th century to the 20th. This is the centre piece of the Summer Palace, west of Peking, which has been extensively rebuilt and modified in different styles, including European baroque.

Left: half-timbered houses were built on a timber framework, free standing on blocks. The walls were then filled in, often with wattle (usually hazel twigs) and daub (plaster made from lime, sand and straw). This house was first built in the late 15th century.

Centuries ago, Europe was covered with forests, but in modern times shipbuilding and the Industrial Revolution have used up so much wood that timber, with its natural insulating properties, is not much used in the British construction industry.

Building, modern

The origins of modern building techniques and materials can be traced back to the latter part of the 18th century, to the birth of the Industrial Revolution. Stone and timber were the dominant materials of construction and the structure of buildings was essentially timber floors supported by stone walls or timber framing. Although a scientific attitude to the design of structures was well established in the European schools of engineering, few builders possessed the necessary technical knowledge to calculate the strength of structures and they relied on intuition and rules of thumb based on mediaeval and Renaissance precedents. Iron was used in small quantities as an auxiliary building material, but its manufacture depended on the use of charcoal, supplies of which were running low. The discovery by Abraham Darby in 1708 that coke could replace charcoal in blast furnaces was of great significance as it meant that iron could be produced in much larger quantities.

The completion in 1779 of the Iron Bridge spanning the river Severn near Coalbrookdale in Shropshire marked the end of the reign of stone and timber as the dominant materials of construction. Its 100 ft (30 m) span, though modest by 20th century standards, was a major achievement and the iron castings for the arch ribs were in two parts only, each one being $5\frac{3}{4}$ tons in weight and approximately 70 ft (21 m) in length.

By the end of the 18th century the use of cast iron instead of wood for columns and beams was developing, but well into the 19th century buildings generally had an enclosing envelope of load-bearing masonry walls.

In the early 19th century the development of railway networks required the large-scale use of iron as a building material but difficulties were encountered with the use of cast iron because of its unreliability in tension, so for beams it was replaced by wrought iron, a much purer material which could be used safely in compression and tension. The industrialization of society brought about great changes in production methods and new concepts of building were evolved: standardization and prefabrication, the introduction of the skeletal steel framework with external glazing replacing masonry walls, the use of calculations and model testing to assess the strength of structures, and a gradual transfer of the focal point of the building industry from the site to the factory. Mid-19th century examples of the extensive use of pre-fabricated units were Joseph Paxton's design for the Crystal Palace in London and Isambard Kingdom Brunel's design for a 1000 bed hospital to be prefabricated in England and shipped out to the Crimea.

Prior to 1870, the principal metallic materials of construction were cast and wrought iron and there was a growing demand for cheap and ductile metal which could be rolled into shapes suitable for construction use. The Bessemer process for making cheap steel which was strong and ductile led to the almost total eclipse of cast and wrought iron by the end of the 19th century. The Forth Bridge (1890), the Eiffel Tower (1899) and Louis Sullivan's famous skyscrapers of the 1890s demonstrated the extensive use of steel in building during this period and the immense strides made in the technique of design and construction in metal from the onset of the Industrial Revolution. The design possibilities of a new material were also being demonstrated—reinforced concrete, which has become the dominant building material of the 20th century.

Modern building materials

Technological developments in the 20th century have produced a new range of materials such as plastics, but perhaps of more importance is the extension of the structural potential of traditional building materials such as timber and brickwork and the development of steel and reinforced concrete, both of which were well established at the end of the 19th century. Concrete and steel have now become the most important building materials.

Reinforced concrete

Concrete is usually made by mixing cement with sand, crushed rock and water. The cement combines chemically with water to form a cement paste around the sand and crushed rock (known as aggregate). The cement paste gradually hardens into a strong material.

Although mixtures like cement were used in Roman times, modern cements were developed from the experiments of Joseph Aspdin, who in 1824 patented a new variety which he produced by burning limestone and clay together in his kitchen stove (Portland cement). Concrete is now used in greater quantities than any other structural material and as a result the manufacture of cement is one of the world's largest industries. The structural properties of concrete are similar to those of masonry—its resistance to compression is high but it has the advantage that it can be poured into moulds to form floor slabs, beams, columns and structures of shell form.

Concrete is weak in tension, and where tension exists it is necessary to introduce reinforcement, generally steel. Reinforced concrete is a composite material in which the concrete resists compression, and the steel reinforcement resists the tension and limits the width of cracks which will occur in the concrete under relatively low load. The introduction of reinforcement made numerous designs possible, and by the end of the 19th century reinforced concrete frameworks were being used for commercial and industrial buildings in Europe and America.

By the 1930s the technique of prestressing concrete had become commercially viable, largely through the pioneering work of the French engineer Eugène Freyssinet (1879–1962). Prestressing consists of inducing a state of compression in the concrete by means of high tensile steel wires so that any tension which occurs because of subsequent loading of the structural member is counteracted by the initial compression. This means that cracking of the concrete can be completely eliminated and thus the possibility of corrosion of the reinforcement by moisture penetration can be avoided. It also allows a structural member to be made up of a series of small sections which

The Guaranty building, Buffalo, New York, designed by Louis Sullivan in 1894, has a steel frame.

Above: New York's Seagram building, by Mies van der Rohe. Below: modern flats under construction in Holland.

can be stressed together. The principle is similar to that of a voussoir (segmental) arch, which is a self-stressing system, but the advantage of prestressing is that it can be applied to a straight member such as a beam.

Steel

By the end of the 19th century steel had become an important construction material. Rolled steel sections were available in quantity and the technique of connecting members by means of rivets or bolts was fully developed. Another important advance made during World War I has now become the standard method of joining steel members, that is, metal arc welding. An arc is struck between a metal rod (the electrode) and the members to be welded. The metal is fused at both ends of the arc and the fused electrode is deposited in the joint in a series of layers until it is filled. This technique is used extensively in the construction of modern steel buildings and bridges, as it produces smaller and more efficient joints than rivets or bolts. The composite action of steel beams and reinforced concrete floor slabs is a further development, in which studs welded to the top flange of steel beams ensure proper interaction between the two materials.

Timber

Although timber is one of the oldest building materials, recent advances in timber engineering have insured that it can take its place among the modern building materials. Of particular importance is the development of adhesives which are insensitive to heat and moisture. This has led to the extensive use of plywood for walls and floors of buildings. Plywood consists of laminates or plies which are peeled from a log and then glued together under

Above: the roof of the new Stock Exchange in London. Because of the soft soil the structure is founded on an enormous concrete raft.

Below: London's new Drury Lane Post Office has welded and bolted sections protected with lead oxide primer.

pressure with the grain in alternate directions. The strength of timber depends on the direction in which it is stressed, its strength at right angles to the grain direction being much less than that parallel to the grain. The use of plies with alternate grain directions ensures a material of uniform strength and dimensional stability. Because of the limited size of commercially available timber sections, the use of adhesives is important in the manufacture of built-up beams of I or box section, and of laminated beams, which consist of a number of layers of relatively thin sections glued together under pressure. Significant developments have also taken place in the techniques of nailing and bolting timber sections and the use of special metal connecting devices. Timber structural elements have the advantage of light weight, the density of timber being about one fifth that of concrete and one sixteenth that of steel.

Ceramics

A large section of the structural ceramic industry is involved with brick manufacturing. Like timber, this traditional building material has undergone extensive rationalization with regard to its manufacture, handling, laying and load-bearing capacity. Brickwork is used as a compressive load-bearing medium in walls and piers.

Detailed studies of the strength of brickwork have made it possible to construct buildings up to eighteen storeys in height with a brick thickness at ground level of 15 inches (38 cm). The stiffening influence of brick infill panels on the strength of steel and concrete frames has been used in recent designs, as has the interaction between wall panels and supporting beams. Design principles similar to those for reinforced concrete have also been established for reinforced brickwork, enabling it to accept both tensile and compressive stresses. Traditionally brickwork has been regarded as a compression material but with the introduction of reinforcement it can be used to make brick walls span large openings as deep beams or cantilevers.

Aluminium

Aluminium in its commercially pure state is a soft, ductile metal but with the addition of alloying elements which increase its strength it is suitable for structural applications. In contrast to steel it is very durable, because of the thin layer of oxide that forms spontaneously on its surface and acts as a barrier against further oxidation. The density of aluminium is roughly one third that of steel but its resistance to deformation is considerably less. The deflection of an aluminium framework will be nearly three times that of an identical framework in steel under the same loads. This and its high cost compared with other building materials has meant that aluminium has found only limited structural application, except in situations where minimum weight is essential. The structural sections available are similar in form to those for steel.

Plastics

Plastics are produced from basic natural products such as coal, air, water and oil by complex chemical processes. Because of their synthetic composition many kinds of plastics can be produced with differing properties but for structural applications they normally require to be reinforced with various types of filler or fibres. One of the commonest types of structural plastic is glass reinforced polyester (GRP), which will withstand high tensile and compressive stresses. This material has the advantage that it can be formed into curved or folded load-bearing elements with relative ease. The stiffness of plastics is low and thus for structural applications this has to be overcome by the proper choice of structural form. The potential of carbon fibre reinforced plastics is still in the development stage, but there have been a number of structural applications of sandwich panels consisting of a strong skin bonded to a low density insulating core. Another interesting application is the use of plastics for air-supported structures (inflatable domes), which are supported by a small pressure difference between the outside and inside maintained by a continuous flow of low pressure air from a fan.

Modern building design

Modern building design is a complex process which involves the interaction of many skills, in particular those of the architect and engineer. Steel frameworks of up to 200 feet (61 m) in height were constructed in the 1890s and typical examples of developments in this form of construction in the period up to 1940 were the Woolworth building of 792 ft (241 m), and the Empire State Building of 1472 ft (449 m). These structures, both in New York, involve the transmission of immense loads to the ground, and the design of foundations is a vital aspect of the building process. Buildings of this size can only be built on ground with a high bearing capacity such as rock. In cities such as London, where the ground is relatively soft, few buildings exceed a height of 400 ft (122 m). Tall buildings are subjected to high lateral loads. Their stability under gale force winds is a major structural problem and the design of the connections between the members is critical.

Design standards with regard to structural stability, durability, fire resistance and environmental control and cost have become much more stringent. These standards, coupled with an increasing awareness of the shortage of national resources, have forced engineers to minimize the quantities of structural materials used in the building process which in turn calls for more refined analytical and constructional techniques.

After World War II the economic climate favoured the use of concrete, reinforced or prestressed, and for high rise buildings the core structure was developed. Tall buildings of the core type are similar in structural form to a tree. Both are essentially large vertical columns rigidly anchored into the ground supporting a series of cantilever projections. The erection of a typical core type building starts with the construction of the foundations. On soft ground piles are commonly used to distribute the load over a wide area to avoid overstressing the soil. The piles are capped with a large concrete slab upon which is built the concrete core. A tower crane is erected at the top of the core and is used to raise large cantilever beams. Ties are hung from these beams to support the floors, which are cast at ground level and then raised. There are a number of examples of this form of construction in London and various cities throughout Europe. In the construction of high and low rise buildings extensive use is now made of prefabricated components and site mechanization. Standardized structural components are now available in concrete, timber and steel and extensive use of these is made in the construction of houses, schools, office buildings and hospitals. In some cases it is possible to prefabricate complete building units including finishes and services. These units can be transported to the site and erected quickly, with the minimum plant and in adverse weather conditions, though not without the cooperation of building trades unions.

Above: the Sydney Opera House, designed by Joern Utzon, has a foundation of 700 concrete piers, 3.3 feet (1m) in diameter and sunk 42 feet (13m) deep.

Below: the American geodesic dome at Expo '67, Montreal, 20 stories high, attracted 9,000,000 people.

47

In modern building the science of statics is used in calculating the stress on the finished structure.

Right: a synthetic material is tested under heavy load to determine its mechanical properties. The internal stresses and strains produced by the application of a load depend on the shape and dimensions of the object as well as the material itself.

Below: a plastic model of a cross section of the French cathedral of Bourges has weights hung on threads running through the small holes. Then it is heated so that it softens and deforms under the load. The distortion is revealed by photographing it through a polarizing filter, which gives rainbow colours to the stressed areas.

Early builders achieved magnificent results by rule of thumb, but they could revise as they went along. Today's construction industry uses new materials and ambitious designs; it works within very tight economic limitations: all the calculations must be done well ahead of time. Large buildings such as schools and office blocks are very expensive, while private dwellings are not only more costly than ever, but are built by businessmen using somebody else's money.

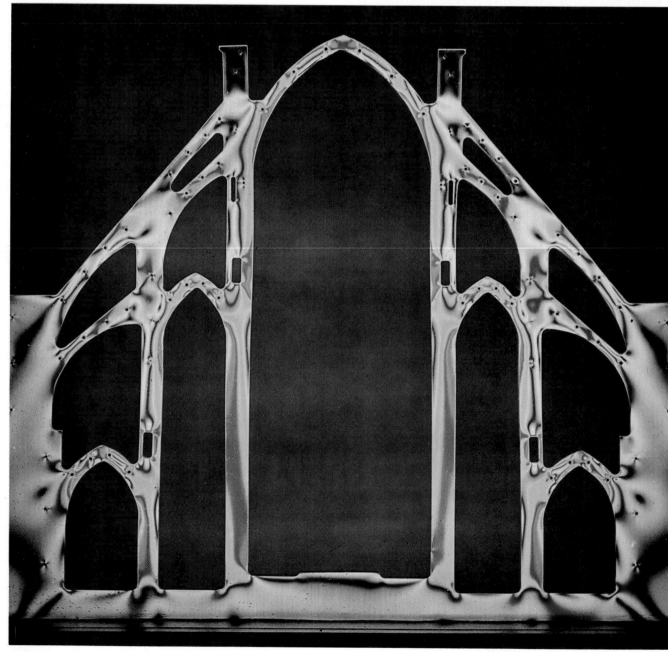

Cathedral construction

The builders of mediaeval cathedrals and other great churches were working in a tradition which can be traced in an unbroken line back to ancient Rome, and from Rome back to Greece. Though classical and mediaeval buildings may look very different, the fundamental methods of design, even the measures used, were directly related.

These methods of design were based on geometry, and though they had little to do with structural engineering they provided the builder with simple solutions to complex problems and produced structures whose dimensions were proportionally related. The builders improved their designs as they went along, and as the large number of surviving buildings shows they were generally successful.

Vault height

One example of this simple geometry is the way that the height of a cathedral's main vault could be decided 'to the square' or 'to the triangle'. If it was to the square, the height of the vault would be the same as the width of the church from wall to wall. To the triangle, the width of the church would be taken as the base of an equilateral triangle, and the height of the vault would be equal to the height of that triangle's apex above its base. Clearly, a church to the square would be higher than a church to the triangle.

Another use of simple geometry was to establish a system of proportional relationships between the various dimensions of the design, and this was very often based on the relationship between the side and diagonal of a square. Once the proportionally related dimensions were determined, the use made of them was very simple. For example, the width of the nave might be made equal to the length of the diagonal of the square whose side had a length equal to the widths of each side aisle. The dimensions of the various elements of the elevation (such as walls and doors) were taken from the dimensions of the floor plan in the same way.

Plans

Very few drawings that might have been used by a master mason survive, and for those that do, none are working plans in the accepted sense. The most important are found in a sketchbook that belonged to Villard de Honnecourt, a French master mason of the first half of the 13th century who visited cathedrals all over Europe in the quest for ideas. He went to Reims while building was in progress, but some of his drawings, while clearly recognizable, show features which were never built. It has been suggested that Villard had access to drawings by the master of Reims, and that he copied them together with details that the mason later discarded.

The acceptance of this deduction, coupled with the numerous alterations in design that can be detected in the fabric of many mediaeval cathedrals, suggests that ultimately the design for such buildings existed only in the master mason's head. Drawings were made to help him decide how certain things ought to be worked out, and to suggest the shape that the final work might take,

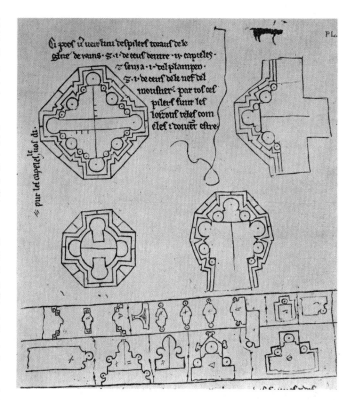

One of de Honnecourt's drawings of Reims Cathedral, made while it was still being built, shows cross sections of piers (above) and mouldings (below) and corresponds closely to reality. Compare to drawing on the next page.

but they were not blueprints.

Masons

In the Middle Ages there was no architect in the modern sense. The man responsible for the design, and usually in charge of the construction, was a mason like the men who worked under him. One of the best kinds of stone used in the Middle Ages was called *Freestone*, a name given to building stone which does not split into layers when worked with a chisel. When freshly quarried it is easy to work, but on exposure to the air it soon begins to harden. Therefore it is usually *dressed* (cut) into cubic blocks, known as *ashlar*, at the quarry. This work was done with a kind of axe—a scappling iron. Not only did it mean that there would be less difficult work at the site, but also that there would be less weight to transport. The men who worked this stone were called Freestone masons. Later this became freemasons, but it has nothing to do with freedom.

Masons were often taken from the quarries when extra men were needed for great projects such as a cathedral. On the site there were various kinds of workers: *layers* were specialists at laying stone upon stone in beds of mortar to build walls; *carvers* used hammers and chisels to carve mouldings, while *ymagers* sculpted figures and foliage.

This design for a chapel at Reims was changed; as built, there are plain buttresses and only two angels.

Foundations

The first stage in building a cathedral involved the preparation of the foundations. The French took this seriously, and often, as at Amiens, dug down to solid rock to ensure a solid base. The English were a little more happy-go-lucky. Some English cathedrals such as Salisbury have no foundations at all; others, York for instance, were built on wooden piles driven down into the soil. Foundations may also take the form of crypts which, being buttressed by the surrounding earth, make very strong structures to build upon.

Plinths

Once the site was cleared and the foundations ready, the next stage was to build the *plinth*, which is the lower part of the walls. The plinth is always clearly distinguishable on the outside of the building, because it is ornamented with horizontal bands of mouldings, and is generally stepped forward from the vertical plane of the walls above, and extends around the bases of the buttresses. From the moment the plinth is built the disposition of most of the major features of the church is fixed. Opposite each buttress a pier will stand inside the building dividing the nave from the side aisle; high above each pier is the *springing point* of the vault, and the job of the buttresses will be to resist the thrust exerted by the vault at that

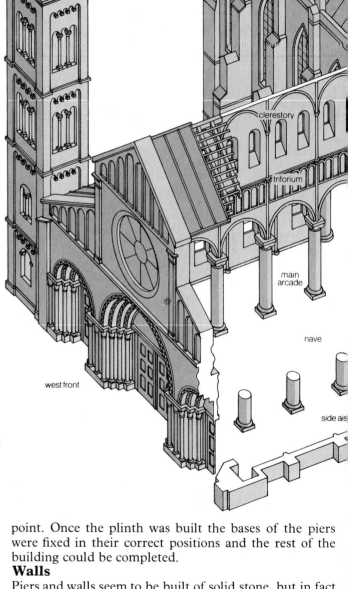

point. Once the plinth was built the bases of the piers were fixed in their correct positions and the rest of the building could be completed.

Walls

Piers and walls seem to be built of solid stone, but in fact they are built from two skins of ashlar with a mass of rubble and mortar in between them. Scaffolding was required as the walls and piers grew higher. It could be of two kinds, one of which was very similar to that used today, but made of wooden poles lashed together and with platforms made from woven willow twigs (*withies*) instead of the planks used today. The other kind of scaffolding employed a simple but ingenious method. As the walls of ashlar were built higher, the masons left holes

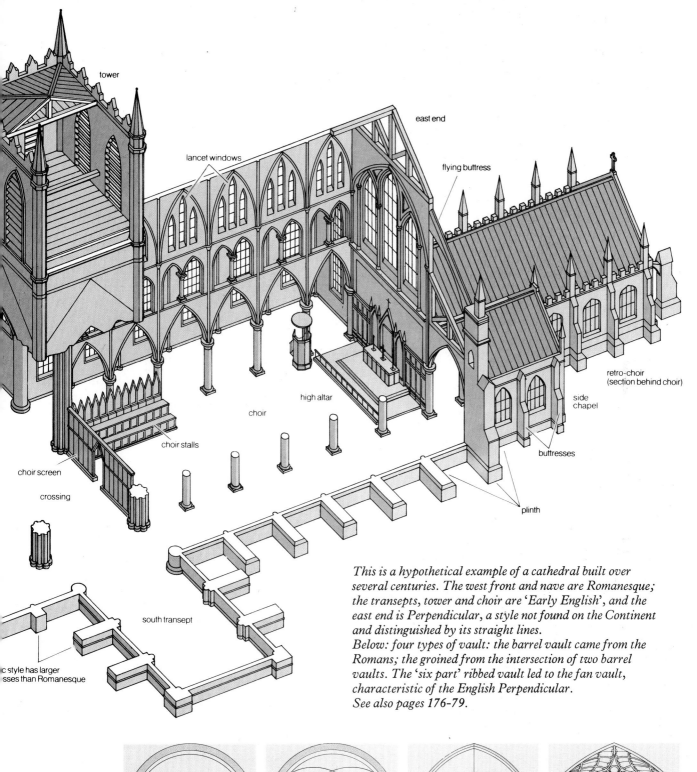

tower

east end

lancet windows

flying buttress

high altar

choir

choir stalls

choir screen

crossing

choir screen

retro-choir
(section behind choir)

side
chapel

buttresses

plinth

south transept

c style has larger
sses than Romanesque

This is a hypothetical example of a cathedral built over several centuries. The west front and nave are Romanesque; the transepts, tower and choir are 'Early English', and the east end is Perpendicular, a style not found on the Continent and distinguished by its straight lines.

Below: four types of vault: the barrel vault came from the Romans; the groined from the intersection of two barrel vaults. The 'six part' ribbed vault led to the fan vault, characteristic of the English Perpendicular.

See also pages 176-79.

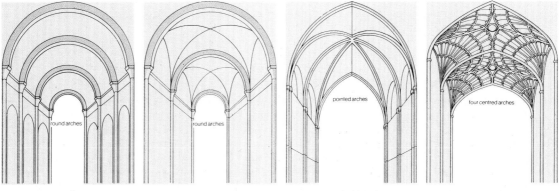

round arches

round arches

pointed arches

four centred arches

barrel vault

groined vault

sexapartite ribbed vault

fan vault

Above: Wells Cathedral shows signs of second thoughts. The strange X-shaped 'strainer' arches were inserted when the tower seemed to be too heavy for the piers.

Below: a 15th century French bible shows Solomon watching the building of the temple: in fact, it is a contemporary building and workmen portrayed.

Above: Cologne Cathedral as it might have been at the end of the Middle Ages, from a drawing of 1820.

Below: York Minster was built on wooden piles. When these began to fail, they had to be replaced with concrete, but this had to be done working from underneath so as not to disturb the fabric of the building.

Above: the Lady Chapel at Westminster Abbey, London, was started in 1503 and took sixteen years to complete. The tomb of Henry VII and his queen in the centre is beneath a beautiful fan-vaulted roof (see p.178).

at regular intervals. Into these holes known as *putlocks* or *putlogs*, they would insert beams which projected from the wall at right angles, and lay the platforms of withies across them. The great advantage of this system was that the same few pieces of scaffolding could be used over and over again. Finally the masons could descend by the same route, either filling in the holes as they went or leaving them open for use in future maintenance work.

Roofing

When the walls had been built the roof was constructed. It was made of timber and often covered with lead or slates. The timber beams were often of oak, and some, called tie beams, had to be long enough to span the width of the nave, and were hoisted by blocks and tackles which were fixed to the top of the walls. Their weight, size, and the precarious position of the men who had to manhandle them into position on top of the walls made this a very dangerous operation. Once there, however, they could be used to support a more effective crane consisting of a giant cylinder, which was turned on the treadmill principle with men walking on the inside, while the lifting rope wound or unwound on the outside. These cranes were important for hoisting the materials for the final major stage—the construction of the vaults.

Vaults

The vaults were formed, a bay at a time, using huge wooden moulds called *centring*. A bay is a rectangular area, defined by the distance between two piers in the same arcade on one side, and on the other, by the width of the nave. The centring was supported on high scaffolding, and took much time and labour to construct. Therefore it was designed to be pushed along from bay to bay, as each vault was constructed. The problem, however, was being able to lower the centring sufficiently to free it from the newly-built vault surface. This could be accomplished by removing wedges from beneath the legs of the centring on top of the scaffold.

The vault surfaces were constructed of roughly squared stones arranged in regular courses. When it was obtainable, *tufa*, a light volcanic stone, was used. The courses were not left visible, but were plastered over. The plaster was often painted in imitation of the very stone courses it concealed. The upper surface of the vault was covered with layer upon layer of cement and rubble, which formed a cohesive shell strong enough for a man to walk over it with little danger of falling through.

Flying buttresses

The lateral thrust of a high vault follows a line which, if continued, would meet the ground some distance beyond the walls of the cathedral. As long as this thrust is contained within a masonry structure, it will not affect the stability of the nave walls. This masonry structure is the 'flying' buttress which consists of two parts, the *flier* and the *buttress*. The flier transmits the thrust to the buttress which, by means of the vertical thrust inherent in its own mass of masonry, is able to deflect the thrust of the high vault to the ground within its own base. Since the deflected thrust is unlikely to be completely vertical, the buttresses are always stepped out as they get lower, to ensure that it will be contained.

Wall thickness

The basic structural difference between *Romanesque* and *Gothic* cathedrals lies in the thickness of their walls. At their springing points vaults exert powerful sideways thrusts, which is unresisted would push the walls outwards and bring the church crashing down. The Romanesque builder counteracted this thrust by constructing walls about 8 ft (2.4 m) thick, which worked as continuous buttresses. Furthermore, he linked these walls together in order to make a succession of rigid box-like constructions.

The emphasis on thick strong walls did not allow for many windows: though desirable for light, they were seen as weaknesses in the structure. Therefore, they were only allowed above the springing points of the vaults in the *clerestory* (clear storey) and in the outer walls of aisles and gallery, and were always kept small.

Gothic cathedrals on the other hand have relatively thin walls (about 2 ft, 60 cm, thick) and large windows, and they dispense with box-like methods of construction. They were able to do this because during the 12th century in France, it was realized that if the thrust from the springing points was completely taken care of by buttressing, then walls would only have to be strong enough to bear the vertical thrust of their own weight and that of the roofs they supported.

Mouldings

Mouldings are one of the principal embellishments of cathedrals. They are carved along ribs underneath arches; on the base of piers and sometimes on the capitals; on the hood moulds which went around the tops of windows; and on string courses running in horizontal bands to divide the surfaces of the building into layers. Mouldings in cross section consist of geometrical designs which were provided by the master mason in the form of wooden

These are some of the many niched statues lining the west front of Exeter cathedral, which was begun in 1275. The statues survived Cromwell and German bombs, but they have suffered from the passage of time. Once they would have been painted in brilliant colours. Occasionally one sees a tomb effigy in Britain restored to look the way it was meant to look; the sight gives the lie to the words 'Dark Ages'.

templates. The carvers transferred these patterns to the stones to be moulded by placing the template at either end of the block of stone, and tracing round its outline with a sharp point. All the stone on one side of the line thus produced would then be carefully chiselled away, leaving the moulding.

Rebuilding

Most of the existing cathedrals were not the first to occupy their sites but simply the most recent replacements of earlier churches which were burned down, fell down, or were even pulled down. Almost inevitably rebuilding began at the east end. This was because the choir was the part of the church used by the clergy; it was where they sat, they wanted it to be splendid and new, and they had the money. The nave belonged to the townspeople and sometimes was left as it was; or if it did follow the choir it often only did so after a considerable interval. At Beauvais, the nave was never built at all; at Bristol it was built only in the late 19th century.

Rebuilding had to interrupt services as little as possible. Sometimes, therefore, the new walls were built around the old, which were taken down later. This made it impossible to make direct measurements between all the points involved in setting out the plan. The result is that the finished work is not always perfectly square or symmetrical.

Construction machinery

Compressors

The Egyptians in 1500 BC used foot-operated blow sacks made from animal stomachs to increase furnace heat or in the smelting of metals. From this developed the bellows, and this remained the main form of air compression until as recently as the 17th and 18th centuries. From this period onwards industrial needs produced more advanced devices to perform a wide variety of functions from tunnelling machines, pneumatic delivery systems, and passenger lifts [elevators], to the operation of machinery for mass production. Even the earliest compressor required a basic energy to power it, muscle power. Once compressed air has been produced and stored it is a source of power, but first of all a prime mover is required to operate the compressor. It may take the form of water power, a gas engine, an electric motor, a petrol [gasoline] or diesel engine, or even a gas turbine.

There are several different basic types of compressor but the simplest example is the bicycle pump. This is the most elementary form of reciprocating piston compressor and takes the form of a leather cup 'piston' moving by hand power up and down a long cylinder, the barrel of the pump. As the handle is pulled out, air is drawn in past the 'piston' until the cylinder is full of air at normal pressure. As the handle is pushed in the leather cup is spread by the resistance of the air in the barrel, which is compressed until it reaches sufficient pressure to pass through the non-return valve at the far end and then into the bicycle tyre. Most cyclists will have noticed that when the air is compressed it becomes hotter, the end of the pump sometimes becoming too hot to touch.

The reciprocating compressor is the most common type for industrial use as a permanent installation powered by an electric motor. It may be mounted on two or four wheels for use on construction sites or road works, and powered by either a petrol or a diesel engine. The main components are a crankcase, crankshaft, piston rod, piston and cylinder, the latter incorporating suction and discharge valves. For pressures up to 5 or 6 bar (72.5 to 87 psi) a simple single stage design is sufficient, but above these pressures two stages of compression are usual, low pressure and high pressure. In the low pressure cylinder the air is compressed from ambient pressure to a pressure equal to the square root of the final pressure. The second stage takes the air to its final pressure. At these ratios power requirement is at the minimum. As mentioned earlier, compression produces heat and in order to reduce temperature problems, which particularly affect lubrication, a cooling stage is normally included between the low and the high pressure cylinders, which may be in the form of a water or air cooled heat exchanger.

Various cylinder arrangements and multiples are available, as numerous as with the internal combustion engine, but often a further cooling stage is required in the form of an *aftercooler*. Air or water cooling can be incorporated according to the size or needs of the installation. The compressed air is normally fed into a

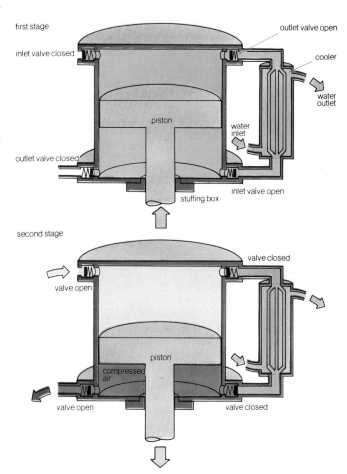

Above: a two-stage piston compressor. Gas is drawn into one cylinder while compression takes place in the other. The gas passes through a water-cooled pipe.

Below: a rotary vane compressor, showing the offset position of the rotor and the sliding vanes.

pressure cylinder or air receiver.

Two other types of compressor design are common, both on the rotary principle. The *screw compressor* comprises one or two pairs (stages) of intermeshing screws which rotate together without touching, taking the air in at one end and compressing it as the screws rotate. This is a very compact design and can operate at a high speed, features well suited for larger air requirements.

The *rotary vane compressor* comes in numerous forms, but basically it comprises a rotor eccentrically placed in a cylinder, the rotor being fitted with a number of vanes which, as the rotor turns, are flung outwards by centrifugal force against the cylinder walls. The sliding in and out of the vanes against the lining of the cylinder forms a seal assisted by lubrication. Because the rotor is offset within the cylinder, air is taken in between the vanes at the point of maximum distance between the rotor and the cylinder wall and is compressed during rotation towards the pressure chamber where the rotor is at its closest to the cylinders.

Up to 5 bar pressure (72.5 psi), rotary vane compressors tend to be single stage, but above this pressure they have to be built in two or more stages. This type of compressor is well balanced but is more suited to lower pressures and, in comparison with reciprocating types, is rather less efficient in relation to the energy consumed.

Applications of compressors are numerous, but in simple terms, stationary compressors are used mainly for industrial purposes, from small garage types for inflating tyres and powering certain tools, to the larger units used for operating air power drills and screwdrivers on assembly lines, for paint spray shops, powering presses, pumping beer and other liquids, and the pneumatic control of advanced production machinery. Portable air compressors are to be seen on most construction and civil engineering sites for operating paving breakers, pumps and so forth. Air tools can be safely used in explosive atmospheres and even under water.

Above left: a compressor undergoing testing in sub-zero conditions, down to −20°C (−4°F).
Above: a rotary pump with rubber impeller. The pump increases the pressure of the fluid by speeding it up.
Right: an irrigation well and 'pump' in Portugal. This is just an arrangement of buckets around a wheel.
A modern pump can be seen as a more efficient version of an ancient idea.

Pumps

A pump is a device for raising a fluid from a lower to a higher level, or for imparting energy to a fluid; although fluids tend to be thought of as being liquid, they also include gases.

In ancient times, irrigation of crops was carried out with the assistance of gravity. This meant that irrigation channels had to be lower than the water source, usually a river. When water had to be raised to irrigate high land or to remove it from a mine, it had to be carried in buckets. Machines were invented such as the *shaduf*, a counterweighted pole with a bucket on one end, and the *saqiya*, a large wheel with buckets around its periphery, to minimize the labour involved, but these methods were finally inadequate.

The earliest account of a pump appears in the literature of the scholars of the Museum at Alexandria, and is credited to Ctesibus' fire engine—a kind of double action pump. The first known pump, however, was used in the Roman Empire after 100 BC. It was a *positive displacement pump*, with a cylinder with a plunger in it and valves at each end. It is also called the Bolsena pump because an almost complete example was found at Bolsena, Italy. The important innovation was the combination of valves and plunger, inventions already known in other technology. The pump had to be precisely constructed out of bronze and was too expensive for most uses; nowadays, however, pumps are the most common machines in use, except for the electric motor.

The Bolsena pump is called a positive displacement pump because with each stroke of the piston a fixed amount of fluid is displaced. The valves are *non-return* valves, so called because the fluid is allowed to flow in one direction only. Such pumps are the *reciprocating* type— that is, one in which the action of a piston, plunger or diaphragm moves a fluid in a pulsated flow. A bicycle pump is a good example; with each stroke of the pump, air is pushed into the tyre and a non-return valve keeps it from coming back out; with the return stroke of the piston air is pulled into the cylinder in front of the piston to be pushed into the tyre on the next stroke. This pumping principle has been the most widely used throughout history, from hand-cranked models to the giant pumps used to drain mines and keep up the water level in canals in the 18th and 19th centuries. A reciprocating pump can also be designed as a double action pump, moving the fluid on both sides of the piston, with the addition of extra valves and a seal where the piston rod enters the cylinder.

Rotary pumps are also positive displacement pumps, but are not reciprocating pumps and have no valves. The fluid comes in an inlet and is pushed by rotating vanes, gears or lobes so that it goes out the outlet. The amount of displacement depends on the clearance between the vanes or the gear teeth.

Positive displacement pumps are more suitable for pumping clean fluids because of the small clearances necessary in the construction; an exception is the *diaphragm* pump, in which the piston or connecting rod is connected to a diaphragm. The back and forth movement of the diaphragm causes the displacement, and because the fluid cannot pass the diaphragm to contact the working parts, this type of pump can be used to move solids in suspension, subject to the ability of the valves to pass the material without becoming clogged.

Peristaltic pumps are rotary pumps without valves in which a flexible tube is continuously squeezed, and are also suitable for pumping a fluid with small solids in suspension, as well as thick liquids. Blood in a heart-lung machine is pumped by a peristaltic pump.

The reciprocating Bolsena-type force pump is still widely used for raising water. Fuel and oil pumps on cars and trucks are usually of the piston and diaphragm type. Hydraulic pumps which operate aircraft controls such as the undercarriage and flaps, jacks to lift cars for repairs, and earth-moving equipment (providing pressures up to 4000 psi or 275 bar) are also positive displacement pumps. They must have some means of relieving pressure on the pump delivery, such as a pressure valve, because excessively high pressures will cause damage to the working parts.

The centrifugal pump began to be developed in the middle of the 19th century. It comprises a wheel with vanes or blades called an *impeller* in a housing or case. The fluid is led into the 'eye' or centre of the impeller through an inlet and the pressure is produced as the fluid is rotated by the impeller at high speed. Extra pressure can be obtained when the high speed liquid is slowed to a more reasonable velocity.

57

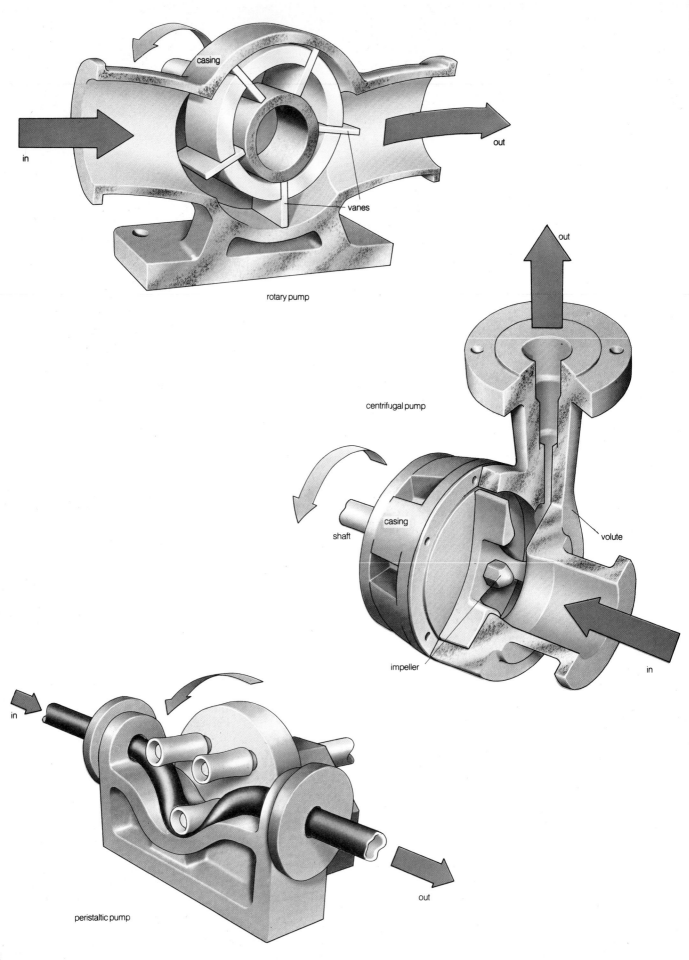

casing

in

out

vanes

rotary pump

centrifugal pump

out

shaft

casing

volute

impeller

in

in

peristaltic pump

out

58

The centrifugal force which provides pressure can easily be understood by imagining a bucket of water swung at arm's length; if it is swung fast enough, centrifugal force will keep the water in the bucket. The pressure recovery from the high velocity is a little more difficult. The shape of the outlet has the effect of changing the low-pressure, high velocity fluid stream into a high pressure, low velocity stream.

The total pressure of a particle of fluid is made up of its *static* pressure, which is what is measured on a pressure gauge, and its *dynamic* pressure, which depends on the speed at which it is moving. The dynamic pressure is the pressure exerted on an object suddenly introduced in front of the moving particle. This dynamic pressure increases as the square of the velocity. It is not possible to convert all the dynamic pressure in a flowing liquid to static pressure, but the pressure recovery can be 0.5 to 0.8 of the dynamic pressure. The simplest method is to slowly increase the delivery channel area (at no greater taper than 8°). This is known as a *diffuser* and is often used on small pumps. On the majority of larger pumps the *volute*, the outer part of the housing, has a cross-sectional area which increases toward the outlet; it is into the volute, which is shaped like a snail's shell, that the fluid is thrown by the impeller. The impeller is mounted on a shaft which is surrounded by a seal where it enters the housing.

Unlike the positive displacement pump, the centrifugal pump requires no pressure relief device because given a certain impeller and a certain rotational speed a predictable maximum pressure is achieved. In effect such a pump is a 'velocity machine' or a *hydrodynamic* pump. Within its performance characteristics it is a very adaptable machine, used for all types of liquids and manufactured from a wide range of materials, from plastic

Above left: the impellers of a high-pressure centrifugal pump used in oil exploration are arranged with the outlet of one leading to the inlet of the next. Pumps (and compressors) arranged in series are capable of impressive performance.

Above: a diaphragm pump, cutaway view. The diaphragm is clamped between the upper and lower castings; it causes the displacement while keeping the fluid from the works.

to bronze and exotic metals such as titanium and tantalum.

Many applications call for the pump to prime itself. If the water level is below the pump inlet, only a positive displacement pump will move the column of air from the suction pipe first, to be followed by normal pumping operation. This is providing the height the water is to be lifted does not exceed the equivalent height of water whose weight produces one atmosphere of pressure. (For fresh water this is 34 feet (10.36 m); it would be impossible to lift higher than this as a perfect vacuum would exist above the water column. In practice water can only be lifted about 28 feet (8.5 m), because of flow losses, vapour pressure of the water, and other factors.)

An ordinary centrifugal pump, when empty, will not remove air from the suction pipe, but once the air is removed by other means, it is 'primed' and pumps normally. The *wet* self priming pump is the most common type and is used extensively in the construction industry. The body of the pump is filled with water, which cannot escape because of a non-return valve in the pump body. Circulation of this water has the effect of removing air from the suction pipe and allows the water to rise into the pump.

Normal centrifugal pumps will not cope with 'snore'; that is, large amounts of air mixed with water. *Vacuum* pumps and air operated *ejectors* are used for priming

these pumps, allowing them to run on snore. Such units are used nowadays in the construction industry where work has to be done below the water seepage level. In the air ejector system, air from a compressor, usually operated by the same motor that operates the pump, is passed through a jet across an opening connected to an air separation chamber on the suction side of the pump. (A pneumatic paint spray works on the same principle.) A vacuum is created in the separation chamber and water is drawn up to the pump which then operates normally. Air which comes up with the water is removed from the water in the separation chamber and blown through the ejector rather than entering the pump.

The pressure produced by a centrifugal pump is roughly proportional to the square of the tip velocity of the impeller, so for a given pressure requirement, the impeller diameter can be large and the speed low or vice versa. The highest rotational speed available from an electric motor (the usual means of driving a centrifugal pump) is 2900 rpm on 50 cycles mains frequency or 3400 rpm on 60 cycles; therefore, when running at these speeds higher pressure can be obtained by using larger diameter pumps.

There are limits, however, one being the strength of the materials used to make the pump; a more important limitation is that the fluid friction produced by the rotation of the impeller affects the horsepower required to drive the pump, and it increases more rapidly above a

Above: fluid power machinery operated by pumps.
Below: pumping concrete at a building site.

certain impeller diameter. For pressures much above 100 psi (6.9 bar), centrifugal pumps are normally built in stages with the impellers arranged in series all mounted on a common shaft, the liquid being led from the output of one stage to the inlet of the next. In recent years, due to experience gained building rocket pumps and multi-stage compressors for jet engines, commercial pumps have appeared running at speeds higher than cycle speeds, enabling a single stage to be used to obtain higher pressures. These pumps are simple, having a gear drive or a belt drive with high speed belts, and represent savings in space and cost. Some of these run at speeds of up to 30,000 rpm producing up to 2000 psi (138 bar).

There are some pumps which do not belong either to the centrifugal or the positive displacement category.

The *air ejector* has already been mentioned as a priming device; it can also be used to lift fluids. The main advantage is that there are no moving parts inside the pump, but efficiency is rather low, at a maximum of about 30%. The most common use of a jet pump, as it is also called, is in conjunction with a centrifugal pump to raise water in excess of 34 feet (10.36 m) and as much as 150 feet (45.7 m).

If large quantities of water are available at low pressure a *hydraulic* ram can be used to lift a small quantity of water from a lower to a higher level. No external power is required; the momentum of the water flow is utilized

when a valve is rapidly closed at regular intervals and high surge pressures are produced. A similar effect can be observed in some domestic water systems when the tap is rapidly closed and a knocking noise or 'water hammer' is heard.

An *air lift* pump is a simple device which can raise water in mines using compressed air, or oil from a well using gas. The air or gas is bubbled into the bottom of a vertical pipe immersed well below the fluid level. The difference in density of the fluid outside the pipe and the gas-fluid mixture inside the pipe causes a flow up the pipe to the surface, depending on pipe length, depth of immersion and the type of bubbles produced.

Liquid metals, such as mercury, are used, for example, as heat-transfer liquids in nuclear reactors. The *electromagnetic* pump is designed to circulate them; a constant magnetic field is passed through the liquid on an axis perpendicular to the desired flow direction. The device works similarly to an ordinary direct current electric motor.

Below: a cutaway view of a centrifugal pump. The motor pulley, left, is mounted on one end of the shaft. The two roller bearings on the shaft can be seen with lubrication reservoir above. The reservoir will provide a constant drip. On the other end of the shaft are the nylon impeller and the inlet. The outlet is at the top.

Piledrivers

A piledriver is a machine used to install piles, which are sheets or columns, usually of steel, driven into the ground to support a building or other structure. It is some 4000 years since the first piledriver was used to install stakes to support primitive lakeside dwellings, and in other periods, in the days of the Romans for instance, piles were used to form the foundations of many known historic buildings still in existence today. The early piledrivers operated on the simple drop hammer principle, dropping a heavy weight (such as a rock held by a rope) on to the pile to drive it into the ground.

Drop hammers used today are made of cast iron and weigh from half a ton to five tons. They are fitted with guides which slide in vertical members to ensure that the hammer remains central on the pile. Although this method is still in use it is rather slow compared with more modern drivers which are powered by air, diesel, or occasionally steam. Some drivers are available which operate by hydraulic or electrical systems.

Piles fall into two main categories. Steel *sheet piling* is used to withstand horizontal forces, which occur in docks and on temporary works. *Bearing piles* are used mainly to build upon, and skyscraper blocks are founded on these except where soft ground makes other types of foundation necessary.

There are two main types of air hammer, *single acting* and *double acting*. The single acting driver is similar in action to a drop hammer, but the heavy falling *ram* is raised by air or steam pressure instead of by a crane.

When the desired stroke (height) has been reached the pressure is released, letting the ram fall and knock the pile into the ground. The rams vary in weight between 2.5 and 15 tons, and the single acting driver is somewhat more efficient than the drop hammer. Single acting hammers are used to install bearing piles.

The double acting air hammer is a somewhat lighter tool used to install sheet piling. It normally operates on compressed air, and has a light ram rather like a piston, which moves up and down within a cylinder, the bottom end of which is the *anvil* that rests on the top of the pile. Air is admitted to the cylinder to lift the ram, which on its upward journey actuates a valve arrangement that cuts off the air supply to the lower part of the cylinder and transfers it to the upper part above the ram. This forces the ram back down the cylinder to strike the anvil, which in turn strikes the pile, and the air is expelled through exhaust ports.

Unlike the air and steam hammers which require an auxiliary source of power (compressors or steam generators), the diesel hammer is a self contained unit. Like the double acting air hammer, the ram is in the form of a piston moving within a cylinder. The hammer is started by lifting the ram with a crane and releasing it when the required height has been reached.

Below left: this pile hammer is designed for offshore oil platform work. It is single acting, steam powered and has a falling weight of 154,000 lbs (70,000 kg).
Below: an early pile driver, just a dropped weight.
Bottom: driving H-section steel piles. Steel piles can be removed by a machine which works in reverse.

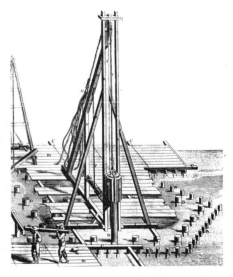

On the downward stroke the ram actuates a cam-operated fuel pump which injects a measured amount of diesel fuel into the combustion chamber at the bottom of the cylinder. The ram continues downwards, closing off the exhaust ports and compressing the trapped air. The ram then strikes the anvil, thus driving the pile, and at the moment of impact the fuel-air mixture (which has been highly compressed in the combustion chamber at the bottom of the cylinder) ignites, and the resulting explosion drives the ram upward to start the cycle again. It is stopped by turning off the fuel.

The single acting hammer is raised by air or steam and then dropped. In the double acting hammer, the lifting air or steam is re-directed behind the hammer when it drops to give extra power.
Right: the ram of a diesel hammer is raised by a crane; when it drops it compresses a fuel mixture which explodes, starting a cycle.

Cranes

In modern industry and commerce, the necessity of lifting and moving heavy loads takes so many different forms that a variety of cranes have been designed to deal with them. Cranes can be mostly divided into two types, with numerous variants of each: *bridge* cranes and *jib* cranes.

Any type of crane may be fitted with one of a number of lifting attachments. Webs, nets, ropes or cables may be attached to the hook. For lifting bulk materials, such as ore, gravel, earth and so on, a clam or grab is used, consisting of two jaws on a hinge which can be opened and shut by the operator. The operator is often responsible for the safety of the lifting operation. There is a mathematical relationship between the available strength of a cable or a rope and the angle at which it is wrapped around the article to be lifted, and the operator may be charged with the responsibility of not lifting the load until it is safely slung.

For indoor applications, such as in machine tool works,

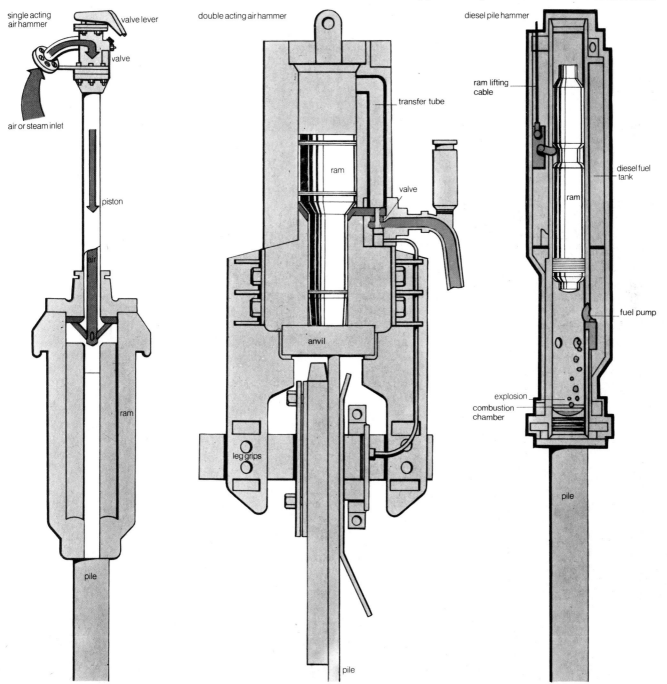

steel mills and other types of factories, a travelling electric bridge crane is used. The hoisting apparatus travels back and forth across the width of the working area on an overhead bridge made of steel girders, while the bridge itself can travel the length of the building on trolley wheels along an elevated framework (*gantry*), which supports it at each end. The gantry is built against the walls of the room, or alternatively is a structural part of the building's framework. Thus the crane can pick up a load and put it down anywhere in the room without taking up any floor space itself.

An electric overhead bridge crane may be operated by means of *pendants*—cables and switches hanging down from the bridge—or by an operator in an overhead cab which travels along at one end of the bridge. A bridge crane may be used almost continuously on a work shift, for example to carry ladles of molten metal in a foundry, or may be used only intermittently, to move large parts as in the erection of heavy machinery.

Gantries for bridge cranes may be built out of doors, as in a stockyard, but more usually the outdoor bridge crane is mounted on legs with wheels which run on rails. In this case it is called a *goliath* crane. Goliath cranes are often designed so that the bridge extends between and beyond the legs, extending the useful operating area; this requires cantilever weights or supports on the legs or on the end of the bridge.

A recent development is the free-path goliath crane, in which the legs are mounted on large pneumatic rubber tyres so that it can be driven around the work area. The wheels can be turned through 90° (with power assisted steering) to make the greatest possible use of the crane through manoeuvrability.

Jib cranes are the most familiar type because they are often mobile and widely used outdoors. The jib is the long boom, nearly always of a lattice construction to save weight, which is *derricked* (raised upwards and lowered outwards to alter the working radius of the machine) and *slewed* (turned) in a circle along with the superstructure. On any crane, most of the lifting is done by the cable which is wound on a drum inside the superstructure. On a jib crane, while the jib can be derricked with the load on it, its main purpose is to make the lifting function available over the work area.

The jib crane, if fully mobile, is usually powered by a diesel engine. The lattice construction of the jib is of great strength, and the weight saving is important because the less the weight of the jib itself the more the machine can lift. The jib consists of an upper and lower section, and most jibs are designed so that extra sections can be added in the middle. The sections are 20 ft (6 m) long and the assembly must be done with the jib on the ground, which requires a lot of room.

Jib cranes are mounted on rubber-tyred wheels for use on outdoor sites where the ground is flat and firm. Where the surface is soft or irregular, the crane will be mounted on a track, like a military tank. The track spreads the weight of the machine over a wider area than wheels so that the pounds of pressure per square inch are much less.

Self-propelled cranes are usually too slow to be driven on the street, so they are usually hauled to the site on flat-bed trailers. In the last decade, the telescoping boom crane mounted on its own vehicle has been developed. It is designed to be driven to the site; upon arrival, the boom is extended hydraulically to its full

On these pages is a drawing of the base and trolley jib of a tower crane. The tower has interchangeable sections to vary the height.

Directly below: a cluster of cranes on a building site. Several kinds of cranes may be needed.

Below right: two tower cranes attached to the building under construction. They are lengthened as the building goes up. Sometimes they are built in the building's lift shaft.

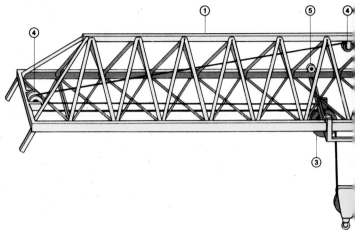

length in a minute or so. This type of extremely mobile crane is convenient and economical for use on a site where there are only a few lifting jobs to be done and where they can all be accomplished within a short period of time. (A construction company has such a large investment in heavy equipment that the more versatile the machine is the lower the actual operating costs are.)

The disadvantages of the telescoping boom crane are that the boom is relatively heavier than the lattice-jib, and the whole machine is smaller. Thus its effective operation is somewhat restricted. To extend its operating range to the maximum, it has hydraulically operated outrigger supports, with hydraulic jacks on their tips. When these are fully extended the crane can reach as far as possible from its centre of gravity for a load without tipping itself over.

In the construction of tall buildings tower cranes are used. The lattice tower supports a horizontal boom

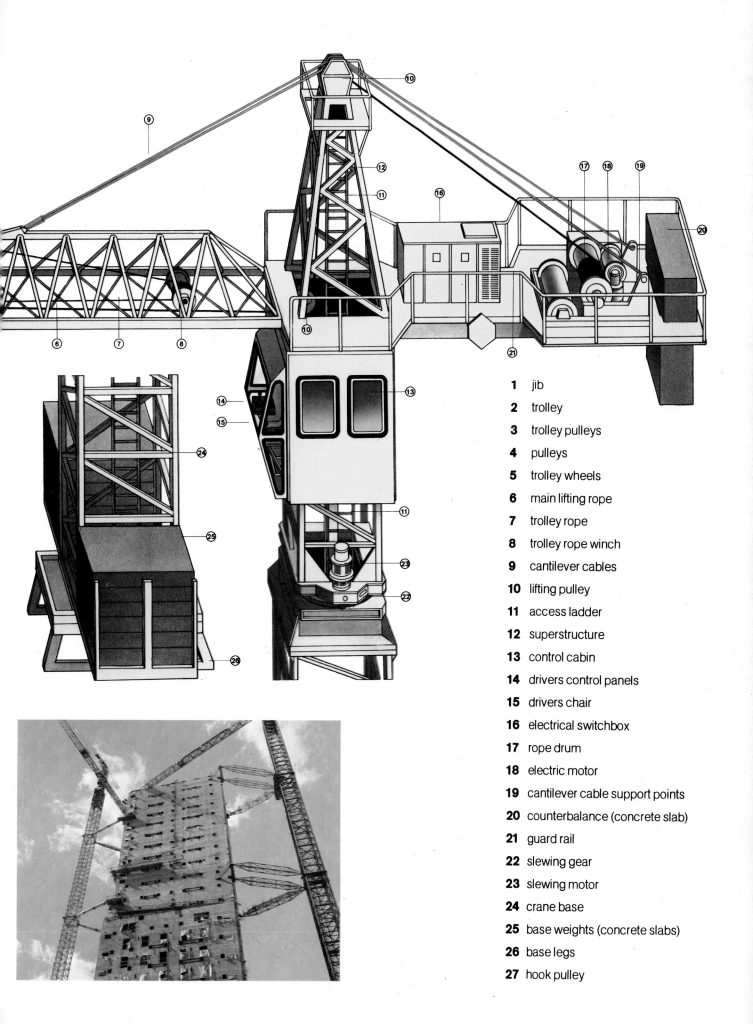

1   jib
2   trolley
3   trolley pulleys
4   pulleys
5   trolley wheels
6   main lifting rope
7   trolley rope
8   trolley rope winch
9   cantilever cables
10 lifting pulley
11 access ladder
12 superstructure
13 control cabin
14 drivers control panels
15 drivers chair
16 electrical switchbox
17 rope drum
18 electric motor
19 cantilever cable support points
20 counterbalance (concrete slab)
21 guard rail
22 slewing gear
23 slewing motor
24 crane base
25 base weights (concrete slabs)
26 base legs
27 hook pulley

which extends on both sides of it in opposite directions. On one end of the boom, usually shorter than the other, a counterweight may be suspended. On the other end the hoisting mechanism runs back and forth on a trolley. The maximum load can be lifted when the trolley is close to the tower, and it diminishes as the trolley moves away from the tower on the boom.

Up to a height of about 200 feet (61 m) the tower can be free standing, mounted in a concrete block or on a ballasted base, but above that height it is fastened to the building at one or more points. An alternative is the climbing tower crane, which is built in what will be a stairwell or a lift [elevator] shaft in the finished building. As the building goes higher the crane is raised along with it.

Nowadays for greater mobility tower cranes are being mounted on ballasted trolleys which run on rails, and even on crawlers and trucks.

Scotch derrick cranes have a rotating mast which is held vertical by two inclined rigid structural backstays attached to the top. Also attached to the top of the mast are the suspension ropes for the low-pivoting derricking jib. The advantage of this type of crane is that the maximum weight can be handled in a wide radius because the legs are secured or weighted far behind the mast. The scotch derrick crane may be fastened to the steel structure of the building and moved higher as the building goes up. Alternatively it may be mounted on structural steel towers called *gabbards*, which may be either static or mounted on a track and powered.

For the rapid loading and unloading of ships the most familiar type of crane used is the *level-luffing dockside crane*. Level-luffing means that the load is neither lifted nor lowered by the jib when the jib is derricked (luffed) in or out. The dockside crane may be mounted on a structure that straddles a railway or a road; in this case it is called a *portal crane*. If it is mounted with one side on the quay and the other side on a gantry at a higher level it is called a *semi-portal crane*.

For unloading of bulk materials from ships the *transporter grab crane* may be used. The load is supported by a trolley which runs on a bridge. This type of bridge crane is sometimes designed to move the load over a considerable distance.

More and more cargo is now being transported, especially by sea, in standard sized containers. Special cranes are installed on the dockside which are similar to the transport grab crane in design and are capable of very rapid and precise handling of these containers.

Under the pressure of increasing trade, and advanced technology in building and industrial processes, cranes are being built nowadays with capabilities which a few years ago were unthought of. A goliath crane used in the building of power stations can lift several hundred tons. A crane used to unload ore from bulk carriers weighs 850 tons and unloads 2000 tons per hour; the grab weighs 21 tons and its contents may weigh up to 24 tons. A crane currently under construction in a shipyard will be 400 feet (121 m) high and will weigh 1000 tons. It will be able to lift 1500 tons.

Hydraulic excavators

These are among the most useful machines available to the construction industry. In capacity they range from small machines suitable for ditch cleaning to the largest which can lift and dump 20 tons of rock at a time, but the typical general purpose machines have a capacity of 1.5 or 2 tons. Many are mounted on pneumatic tyres [tires], but most run on crawler type tracks. The smaller designs are often based on derivatives of agricultural tractors, but the majority are purpose-built from the ground up.

The usual power source of a hydraulic excavator is a modified automotive diesel engine, but with high pressure oil pumps fitted in place of the usual clutch-torque converter-gearbox transmission assembly. The boom movements are made by double-acting piston and cylinder rams, and hydraulic motors power the slewing motion and drive the crawler tracks. Wheeled machines usually have a geared drive for travelling.

The hydraulic systems are controlled by valves fitted into the oil lines that connect the pumps with the rams, and the driver operates the valves by means of levers and pedals in the cab. The larger machines usually have servomechanisms to reduce the effort required from the driver. Oil pressures vary from one design to another, but typically range from 105 to 340 bar (1500 to 4850 psi).

The common factor shared by all hydraulic excavators is a sturdily built boom with three or more hinged joints, which carries the digging implement. In all crawler and wheeled machines, other than the tractor-based designs, this boom is mounted on a *slewing platform* which also carries the operator's cabin and the engine and its ancillary components. The slewing platform is mounted on an undercarriage which comprises the running gear, and in all modern designs it is able to revolve continuously in either direction.

Above left: looking along the boom of a tower crane. As the trolley moves further along the boom, it can lift only correspondingly smaller weights.
Left: a drag shovel crane at an opencast coal mine. This machine grabs 200 tons in one bite.
Opposite page: this wheeled excavator has a backhoe at the rear and a loading shovel at the front.

Tractor-based excavators have the digging attachment mounted at the rear, and a common adaptation permits the base of the boom to be moved bodily sideways. Thus although the boom swing is limited to only about half a circle, the machines can still work effectively in confined spaces. This type of excavator usually has a hydraulically operated loading shovel attached to its front end.

The hydraulic excavator began as essentially a trenching tool. In this form it is generally known as a *backhoe*, and it digs by pressing the leading edge of its bucket into the soil, then rotating and thus filling the bucket by means of the boom rams. The load is then raised by the boom, which is swung away from the excavation by the slewing motion. The load is dumped by contra-rotating the bucket.

Among the numerous attachments available to increase the versatility of the machines are special purpose buckets for light or dense materials, ditch cleaning and weed cutting; ripper teeth for breaking layered rock and paving; 'clamshell' buckets for sinking circular or square shafts; extension pieces to lengthen booms; and magnets and grabs for scrap metal handling.

The most important adaptation of the base machine, however, is the *face shovel*, in which the digging action is reversed so that the bucket is loaded by pushing forward rather than by being dragged back. This may be achieved by rearranging the standard boom components, but most machines use face shovel boom assemblies which are interchangeable as a whole with the backhoe assembly. Face shovels are usually used for hard, heavy digging, as for example in strip mining and quarry work.

The backhoe is invariably arranged to provide the maximum reach, digging depth and dumping height commensurate with machine stability and a reasonable working life. Wheeled machines often have hydraulically operated stabilizing arms to support the machine on sloping or uneven ground. Face shovels, however, stand close to their work and are rarely able to dig below ground level. Typically, they shovel material from the foot of a quarry face, slew through a quarter or half circle, and empty the material into a truck.

Since the immediate post-war years, when hydraulics technology was first seriously applied to earthmoving machines, the hydraulic excavator has steadily superseded the cable (wire rope) operated machines (such as the dragline excavator) for most small or medium sized tasks, but as yet it cannot compete with the larger rope operated equipment.

Bulldozer

During the 1920s the crawler tractor became a reliable and economic machine, and the bulldozer was developed from it. It is a modified crawler tractor weighing from 6 to 40 tons with an engine producing from 50 to 700 horsepower, with a pushing blade across the front, wider than the tractor and vertically curved so that the load is rolled along. The blade can be raised considerably to clear a load and lowered a little to cut the ground surface. On older bulldozers it was operated by winch and cable, but now it usually works by hydraulic power.

Development has had little effect on the bulldozer's appearance. Petrol [gasoline] engines have been replaced by diesel, and automatic or semi-automatic transmissions have replaced manually operated crash gearboxes. Other improvements have led to track assemblies capable of running reliably for long periods in adverse situations.

Far left: this tractor-type excavator has hydraulic stabilizing arms which can be adjusted for sloping or uneven ground.

Near left: a crawler-mounted slewing platform excavator.

Below left: the familiar bulldozer being used to move loose rock. This one is an International Harvester TD 25C model.

Right: a small dumper truck of the type used on building and demolition sites. It has front wheel drive and rear wheel steering. This model has a capacity of 1.5 tons.

Below right: a large bulldozer with a 'ripper' attachment at the rear. This is used like a plough to break up hard ground; then the blade handles it in the normal way.

The ground pressure exerted by bulldozers is rarely more than that of a man's foot and is considerably less when they are mounted on extra-wide tracks. Sometimes the working blade is mounted so that the load is discharged to one side, and the machine is then called an angledozer.

Bulldozers are in wide use throughout the world, mainly for clearing undergrowth and rough-forming flat surfaces, but many bull- and angledozers spend their time hauling other appliances such as compaction rollers or 'box' scrapers which lift spoil in thin layers and spread it evenly over large areas.

A successful derivation of the dozer is the crawler loader, which was originally a mechanical shovel and has now become a multipurpose tool capable of performing other tasks as well as bulldozing.

The low maximum speed of a tracked machine—6 mph (10 km/h)—coupled with great advances in giant pneumatic tyre technology, has led to special purpose machines which are often capable of outperforming the dozer. Striking examples are motor scrapers, which can lift, carry, and spread up to 50 tons of suitable material, and have a top speed of 30 mph (50 km/h). These are now in common use on road formations and similar earthworks. In order to minimize the damage caused by wheelspin, a bulldozer is often used to push motor scrapers into the cut.

Advances in power transmission hydraulics have also led to improvements in machines that are only distantly

related to the dozer. Wheeled loading shovels or hydraulic excavators working with 50 ton capacity dump trucks are frequently used to move large quantities of rock and earth.

The usefulness of the dozing principle has given rise to wheeled versions, geared to move considerable loads but still able to run fast during the light part of an operating cycle. These machines are almost invariably used to shift light materials such as coal or wood chips.

Dumpers

Very few kinds of earthmoving machinery are adapted to moving material more than a few paces, and most depend on some kind of vehicle to move the spoil from one place to another. Often standard roadgoing 'dump' trucks are used, particularly when public highways have to be traversed, although a common modification made to improve traction on poor ground is to transmit the drive to all of the wheels. Such vehicles are, however, not sufficiently robust to withstand the conditions encountered in earthmoving applications and, since reliability is usually more important than first cost, most transport is specially built and very sturdy. The dumping mechanism is usually powered either by screw jacks driven by the vehicle engine, or by hydraulic rams.

Vehicle sizes range from about 20 tons payload up to 200 tons, although 100 tons is accepted as being the maximum capacity currently practicable on two axles, while the economics of bulk earthmoving tend to favour 50-ton vehicles as the dumper capacity and size must be matched to the excavating and loading equipment.

At the other end of the scale there is a wide range of small dump trucks in use on construction and demolition sites, working as replacements for hand barrows and carts. The skip is at the front, and hinged so that it can tip forward to discharge its load when the securing latch is released. The skip is designed so that the weight of the load acts in front of the pivots, so that when the latch is released the skip will tip with little or no assistance.

Above: a purpose-built heavy duty dump truck, with four wheels per axle. A wheeled loader can be seen in the background.

Right: a large motor scraper. A similar machine, the grader, has a bulldozer-type blade instead of the scraper box.

Right top: the largest dragline excavator in Europe, at an opencast coal mine in Northumberland, England. Called 'Big Geordie', it weighs 3000 tons; the bucket can hold about 100 tons of earth, and is carried on a boom 265ft (81m) long.

Small dumpers are usually powered by a small air-cooled diesel engine, which drives the front wheels, and the steering acts on the rear wheels.

Continuous working

The bucket-wheel excavator and the belt conveyor have the advantage of operating continuously, unlike other machines which spend only a fraction of their time actually digging and carrying. The search for other means of continuous working has, in a limited way, been very successful, and an established way of excavating and moving sand, ballast, clay and other materials is to wash it free with jets of water and then pump the resulting slurry through large pipelines. This is similar to the methods used in the extraction of china clay, and huge quantities can be moved over considerable distances in this way. A great deal of land reclamation is carried out in this way, while essentially similar techniques are used by suction dredgers to deepen waterways.

Attempts to replace conventional hard ground digging machinery with other methods have so far been largely unsuccessful. The use of controlled nuclear explosions to produce major excavations has been suggested but is as yet untried, although great quantities of conventional explosives are in daily use. Ultrasonics (high frequency sound vibrations) are a possible means of breaking up ground and rocks, but an effective way of using this technique has not yet been found.

Proposals to use air-cushion type carriers in place of wheeled or tracked vehicles have so far found little

response, as the advantages of dispensing with expensive gears, tyres, tracks and transmissions are outweighed by the need to keep machines as versatile as possible.

Scrapers

The development of large, reliable tyres has aided the development of motor scrapers, machines which are used wherever ground conditions are suitable, for moving quantities of soil over moderate distances. Essentially the motor scraper is a big open-fronted box, mounted on wheels and driven by a large diesel engine, and whose leading edge can be pushed a little way into the ground. As the vehicle moves forward the box fills, the load being pushed out at the discharge point by a moving bulkhead, which gives a controlled discharge as the machine continues to move forward.

Tyre life is a dominant factor in the economics of motor scraper operation, and a recent development seeks to lessen the number of buckets mounted radially around the rim of a large wheel. This wheel, rotated under power, is carried at the end of a large boom that can be raised, lowered and swung so that a continuous cutting action can be maintained. Spoil is deposited by the buckets as they invert on to a system of belt conveyers that carry it away to some form of transport.

Most bucket-wheel excavators are huge—some are among the largest of self-propelled land machines—and they are well established in such applications as opencast coal mining where, for all practical purposes, the quantities of material to be moved at a particular site can be considered limitless. At the other end of the scale, work is going on to develop much smaller machines which would be suitable for ordinary public works projects. Some extremely large dragline excavators are also in use in opencast mines and quarries.

Loaders

Crawler-mounted loaders, commonly confused with the outwardly similar bulldozers although their functions are usually quite different, are among the most generally useful earthmoving machines. The crawler loader is particularly suited to work on demolition and rough site clearance, where its ability to manoeuvre in confined surroundings outweighs its disadvantage of a limited top speed. The main difference between crawler loaders and bulldozers is that the former have a loading bucket at the front instead of a blade, but they both have the advantage of being able to work on broken rock, concrete, or other jagged debris which would quickly ruin pneumatic tyres.

Wheeled loaders are used where ground conditions are good and the bucket can be filled without wheelspin. They are usually fast and capable of making rapid turns, and can run as quickly in reverse as in the forward direction. They are often found in materials handling applications, and the largest and most robust models are able to provide satisfactory performance in quite arduous quarrying and similar applications, due mainly to improvements in tyre design.

Above: the cutting wheel of a bucket-wheel excavator used for scooping up crushed ore at a works in Western Australia, similar to the machine pictured on page 127.

Opposite page: dredgers are important not only for widening and deepening waterways but also in alluvial mining. There are grab dredgers and suction dredgers; this one is a stationary chain-and-bucket type, which is somewhat manoeuvrable by means of the anchor chains.

Construction materials

Bricks

Bricks are widely used as a building material, especially in countries that are particularly rich in different varieties of clay and can therefore produce a wide range of types, strengths and colours. Bricks are joined together with mortar to form walls (including load-bearing ones), veneers, arches, piers, foundations and various other structures. The bricks used to line furnaces and fireplaces are made from fireclay, which is *refractory*—they can withstand very high temperatures without softening. Well-made bricks are waterproof, fireproof, do not rot and are resistant to fungus attack.

Clay is a hydrated silicate of alumina often containing small amounts of impurities such as iron oxide. The clays used in brick manufacture must be sufficiently fine grained and plastic when mixed with water. Firing changes the chemical and physical structure of the clay because the heat fuses the clay particles together into a hard cohesive mass which is virtually indestructible.

The earliest bricks were made of mud and straw. They were formed at first by hand and later in a mould, and sun-dried. Once the art of pottery was discovered, however, bricks were made from clay and burnt (or *fired*) in kilns. The earliest known clay bricks were found on sites in Mesopotamia dating from about 2500 BC.

In Europe, bricks were used widely by the Romans. The fall of the Empire caused a temporary loss of the art of brickmaking, but it was revived during the Romanesque and Gothic periods. It was re-established in England in the 13th century, and led to the eventual creation of such masterpieces of brick construction as Hampton Court in the early 16th century. At this time bricks were still hand made, but in 1619 the first patent was taken out for a clay working machine. The Great Fire of London in 1666 changed London from a timber-built to a brick-built city.

Mechanization in brick-making did not begin until the mid-19th century. Before that, bricks had always been fired in intermittent kilns. In this method, 'green' (moulded and partially dried) bricks would be loaded into the kiln and fired. After the required time the fire was put out; the kiln was opened and the bricks were allowed to cool before they were removed. This process was repeated continually. In 1858 a man called Hoffman introduced the continuous kiln, which was circular and contained from 10 to 20 separate chambers. The principle is that the fire is led from chamber to chamber in turn so that batches of bricks can be loaded, fired, cooled and removed in permanent rotation. Today the vast majority of bricks are fired in continuous kilns which are usually arranged in two parallel end-connected sections rather than in a circle.

Facing bricks are used for visual effect, combining attractive appearance and colour with good resistance to

Below left: soft and fine-textured clays are used for brick making for the extrusion process. Water is added to the clay to produce the right degree of plasticity; after fine grinding the clay goes to the pug mill where it is extruded into columns. Below: it is cut by wires.

exposure. Engineering bricks have great strength which makes them suitable for civil engineering projects such as sewers, engine pits and power stations. Common bricks ('commons') are used for general construction where appearance is not of prime consideration.

Most additional names are purely descriptive, sometimes inaccurately so. Flettons are bricks made from the clay of the Bedford and Peterborough area in England.

They have sharp edges and a deep *frog* (the indentation in the top and sometimes the bottom of a brick for keying the mortar at joints). *Stocks* sometimes means bricks that are hand made (in a 'stock'), but often means the yellow London stock brick—which is not made from London clay. *Specials* are bricks of non-standard size or shape that may be required for special contracts. There are names that refer to manufacturing methods, such as *hand*

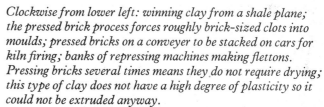

Clockwise from lower left: winning clay from a shale plane; the pressed brick process forces roughly brick-sized clots into moulds; pressed bricks on a conveyer to be stacked on cars for kiln firing; banks of repressing machines making flettons. Pressing bricks several times means they do not require drying; this type of clay does not have a high degree of plasticity so it could not be extruded anyway.

The main firing zone above a tunnel kiln showing gas burners followed by oil burners. The whole cycle of firing bricks takes four or five days.

made, *wire-cut* and so on, and others, such as brindled and sandfaced, that refer to a brick's colour or texture.

The first stage of manufacture consists of mining the clay—it is usually mechanically excavated from seams between one and a hundred or more metres thick. At the factory, raw clay is crushed and ground to a suitable grain size, and watered to produce a workable consistency.

The next stage is shaping the bricks, and there are two basic ways of doing this: either by moulding (sometimes referred to as pressing) or by extruding.

The *stiff-plastic* moulding process is used for clays and shales which do not easily develop high plasticity ('stretchability'). Here the clay is formed to a dough-like consistency (15 to 20% moisture content) and emerges from a pugmill as a rough 'clot', which is then dropped into the mould and pressed. In the *semi-dry* moulding process, used for clays with low natural plasticity, the clay dust is placed while still comparatively dry (as low as 8% moisture) directly into the mould without first being formed into a clot.

Clays with good plasticity, again with 15 to 20% moisture, are *extruded* into continuous columns which are then wire-cut into correct brick lengths, rather like wire-cutting a rectangular cheese into blocks. These may then be pressed to improve surface texture, and must be dried before firing.

The *soft mud* or *slop moulded* process used for making stocks and multicoloured bricks is based on clay combined with lime or chalk in a very wet mix with a moisture content of at least 24%. The bricks have to dry over a long period, and are either extruded, wire-cut and pressed into sanded moulds, or else pressed directly into boxes of three to six moulds each.

Plastic clays of high quality are used for hand-made bricks. This is a common manufacturing method in underdeveloped countries and is also used sometimes for surface effects that cannot be readily machine produced. By using different moulds, some in several parts, many combinations of shapes and sizes can be produced. The individual wooden mould is first wetted to create an adhesive film, then sanded to impart a pleasing texture to the final surface. Next the craftsman strikes off the amount of clay he needs, rolls it in sand, and 'throws' it into the mould—a highly skilled job, since the clay must fill it completely and reach all the awkward corners. Excess clay is bowed off to level the exposed surface, which is then held by a wetted flat piece of wood to keep it smooth while the brick is removed from the mould.

Firing induces chemical and physical changes in the clay, so by varying such factors as temperature and firing time, the properties and appearance of the finished bricks can be accurately predetermined. Colour, for example, is caused by the interaction of iron oxides with various constituents in the clay, and can be modified by varying the amount of air admitted to the kiln at later stages of the firing. It can also be changed by adding a small amount of manganese dioxide, which gives a brown colour.

Firing temperatures of 950° to 1200°C (1762°–2432°F), depending on the type of clay, are produced in kilns. The fuel, such as coal, oil or liquefied petroleum gas (LPG) is charged through openings in the charging floor and burns through shafts which have the green bricks stacked around them. An alternative is provided by car tunnel kilns; here, bricks are placed on the end car of a line at one end of a straight tunnel, and then passed slowly through the drying, preheating, firing and cooling zones before emerging at the other end. Such kilns, typically 400 ft (130 m) long, are becoming more widely used. Firing time may be from four to ten days depending on the type of brick and the kiln. Hollow clay blocks are produced by extrusion, dried in tunnel or chamber driers, and fired in zigzag Hoffman or tunnel kilns. They have

The durability of Roman mortar can be seen in this wall (left) in what is now St. Albans, England. By contrast, mediaeval mixtures of sand and lime crumbled (right).

Above right: concrete blocks are bedded in a modern plasticized cement and sand mortar. The plasticizer works by trapping air so the binder paste fills gaps.

textured faces and are used for external walls on factory or farm buildings, or as the inner leaf of cavity walls.

Calcium silicate or sand-lime bricks, of pleasing light colours and popular for mass housing schemes, are made from silica sand, lime and water. The ingredients are mixed together and mechanically compacted under high pressure into brick shapes. These are subsequently pro cessed by high pressure steam in an autoclave.

Mortar

The kiln-burned bricks used to build the ziggurats at Ur, about 2000 BC, were bonded with bitumen, and it was not until the first century BC that durable building mortars were developed through the expertise of the Romans. In his writings Vitruvius gave detailed accounts of the burning and slaking (addition of water) of lime and the proportions in which it should be mixed with sand for plastering and stucco, as well as the preparation and use of hydraulic cement (*pozzolana*), which was incorporated into both mortar and concrete mixtures.

The mortar used in northern Europe, during the mediaeval and later periods, consisted of lime and sand, but because lime burning and the proportions of lime in the mixture were not as scientifically controlled as with the Romans, much of the mortar in stonework from this period is in a dangerous crumbly state.

Today mortars are used in a variety of building jobs including plastering, rendering and masonry bonding, providing weatherproof joints and surfaces. Basically they are putty-like mixtures of cement, sand and water; often they include lime or a mortar plasticizer to improve the workability. The proportions, however, in a mixture depend on the type of construction work involved. Lime helps to provide a 'fatty' consistency so that the mortar both clings and spreads; it also helps the mortar to retain

water so that it does not set too rapidly but gradually stiffens as the water is lost through evaporation and absorption into the adjoining masonry. The lime also assists the mortar to absorb local strains without accumulation of movement. Plasticizers are useful with low cement and sand mixtures and work by enclosing air in the mixture thereby helping the binder paste to fill up any gaps in the sand.

Coloured mortars are obtained by grinding between 5 and 10 % of a pigment together with the cement.

The type of mortar most frequently used in plaster is cement, lime and sand in the proportions (by volume) 1:1:6 or 1:2:9; each is suitable as an undercoat on most types of background and provide strong hard surfaces suitable for damp conditions. They may also be textured and are applied with a wooden float. Alternatively, cement and sand plasters having similar qualities may be used.

For rendering external surfaces two coats are generally needed and often finishes such as pebbledash, roughcast or texturing are incorporated where a wall is likely to be very exposed to bad weather. As with plastering, succeeding coats should be weaker than the preceding ones. Proportions of cement, lime and sand may vary from $1:0\text{–}\frac{1}{4}:3$ to $1:2:9$ depending on whether the rendering is an undercoat or a final coat and on the finish.

In bricklaying a variety of mortar mixtures are used depending on the type of brickwork such as outside free-standing walls, window sills and inner walls. Mortar mixtures for cement, lime and sand formulations range from $1:0\text{–}\frac{1}{4}:3$ to $1:3:12$. Better frost resistance, however, is provided by mortars consisting of cement and sand with a plasticizer. Batches of mortar can be proportioned by weight or frequently by volume and are normally mixed by hand in small quantities to be used quickly.

Cement

The earliest builders continually tried to find compounds which would hold individual bricks or stones together and spread their load. These compounds are known as mortars and usually consist of an inert substance such as sand, a binder and water. The binder is called cement. Most cements currently used in construction are *hydraulic* cements, that is they react chemically with water to set and harden.

The Egyptians used a mortar which contained gypsum (a calcium sulphate) for pyramid building. Slaked lime (calcium hydroxide) was used by the Greeks and Romans as a mortar in many of their buildings. This is made by heating limestone, which is mostly calcium carbonate to form lime (calcium oxide). The lime is then *slaked*, that is reacted with water. Unfortunately mortar made with slaked lime tends to crack and crumble when exposed to weather, so both the Greeks and Romans found a much more satisfactory, truly hydraulic, cement called *pozzolana*, made from finely ground lime, sand and volcanic material which was found in particular near the Italian town, Pozzuoli.

When water was added it set hard. This type of cement was used in building both the Pantheon and the Colosseum, and remained in use until the late 18th century, although the quality deteriorated in the period after the Romans. John Smeaton, who was commissioned to rebuild the Eddystone lighthouse off the coast of Cornwall, England, experimented with hydraulic limes made by heating limestone and clay to eliminate water and carbon dioxide, and found a product which was superior to pozzolanic cement for underwater use.

In 1824, Joseph Aspdin took out a patent in England on a process for making Portland cement. The name,

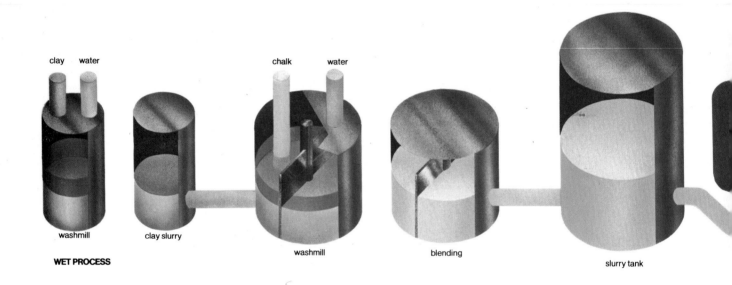

WET PROCESS

clay water washmill clay slurry chalk water washmill blending slurry tank

Flow charts showing the manufacture of cement. The wet process is used where the raw materials are chalk and clay. Water is added to the chalk to form a slurry, which is blended with the clay and pumped to a rotary kiln where the water is boiled off and it is converted to a clinker at a temperature of over 1000°C (1800°F). The dry process is used for harder raw materials, such as limestone and shale; these are ground into a raw meal, pre-heated and passed through the kiln. The clinker is then ground to produce cement.

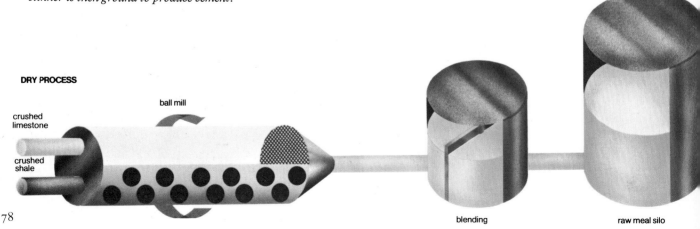

DRY PROCESS

crushed limestone crushed shale ball mill blending raw meal silo

which is not a brand name, was used because of a similarity in colour between the set cement and Portland stone. This process used a higher calcining temperature than had been used before and produced a cement with a much improved strength.

Cement is relatively cheap. It is often manufactured near the quarries which supply the raw materials; it is more economic to move and handle the finished product than the coarse starting material. The raw materials are chalk or limestone, the source of calcium carbonate, and clay or shale, the source of aluminium silicates. Other materials, such as blast furnace slag and even iron ore, are used if they are readily available nearby. The composition of the input to the kiln is carefully monitored, as the properties of the end product can vary considerably with different components. The ground raw materials are mixed together and enter the top of a cylindrical rotating kiln. Where the raw materials have a high moisture content a 'wet' process is used and the raw materials are made into a slurry which is fed to the kiln. This is slightly inclined and revolves slowly. Heat, which is provided by powdered coal, oil or gas, is applied at the lower end. The kiln can be 400 ft (120 m) long. At the operating temperature of about 1500°C (2700°F) the mixture is partially fused to form lumps of cement clinker. The clinker is then cooled and can be conveniently stored as reserve stock even in the open, whereas finished cement can only be kept for relatively short periods in carefully controlled atmospheric conditions. To prepare powdered cement a small quantity, about 2%, of gypsum is added to the clinker and the mixture is ground into the very fine powder that is known as Portland cement. At each stage in the cement manufacture the composition is carefully monitored to ensure that the final product has the desired

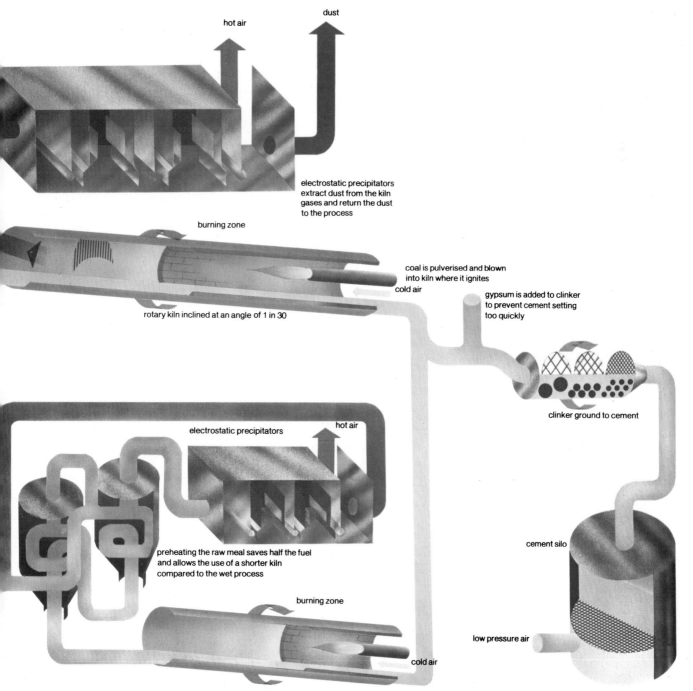

hot air

dust

electrostatic precipitators extract dust from the kiln gases and return the dust to the process

burning zone

coal is pulverised and blown into kiln where it ignites

cold air

gypsum is added to clinker to prevent cement setting too quickly

rotary kiln inclined at an angle of 1 in 30

clinker ground to cement

electrostatic precipitators

hot air

preheating the raw meal saves half the fuel and allows the use of a shorter kiln compared to the wet process

cement silo

burning zone

low pressure air

cold air

Adjusting the feed of the crushed limestone, used in the dry process, to the grinding mill where it is ground into fine powder.

properties.

As cement is a fine dry powder it will flow like a liquid when under pressure. This property is used to deliver cement in bulk, where it is pumped into pressure vehicles by the manufacturer and then pumped out into the customers' silos. It is considerably cheaper to deliver cement in bulk than to bag it and consequently about 75% is delivered this way in Britain, and an even higher proportion in the United States. As cement is such a fine powder it is important to control the dust level. The dust is controlled by careful handling and by the use of electrostatic precipitators and other filtration devices. The electrostatic precipitators are a way of making fine dust particles settle out by electrically charging them and applying an opposite charge at a collecting point.

Cement is a complex combination of four main constituents, which are made in a dry form by partial fusion. When water is added to this mixture it reacts to form an interlocking mass of great strength and hardness.

This dryness of the constituents is chemically termed *anhydrous* which means without any combined water or *water of crystallization*. The four main constituents are tricalcium silicate, tricalcium aluminate, dicalcium silicate, and tetracalcium aluminoferrite.

The addition of water to these anhydrous compounds causes two possible chemical reactions to occur. One, hydration, is the addition of water molecules to the crystal lattice, often changing its structure. The other, *hydrolysis*, is the formation of an acid and a base by the

chemical reaction of water on a salt. In many cases these reactions release much heat. The result is the formation of a complex interlocking matrix of long needle-like crystals which gives the material its great strength. The chemical composition of this matrix is not fully understood but the strength is probably the result of silicon to oxygen bonds.

A small amount of gypsum is added to the cement powder to prevent it from setting too rapidly when mixed with water. Cement paste ages over a period of several years and increases in hardness and crystalline character.

The production of cement from limestone (calcium carbonate) and clay (aluminium silicate) involves the partial fusing of calcium oxide, formed by decomposition of calcium carbonate, and the aluminium silicate to give the mixture of complex silicates that is cement.

Ordinary Portland cement is half tricalcium silicate, a quarter dicalcium silicate and roughly equal amounts of tricalcium aluminate and tetracalcium aluminoferrite. If the cement is to be exposed to acidic action then the proportion of tricalcium aluminate is reduced. A cement which has a low heat of setting is required where considerable thicknesses are used; otherwise the structure can crack through overheating. Both the tricalcium silicate and aluminate react with water, liberating heat, so limits on these are made when low-heat is required.

Cement is mainly used in making concrete, which is a result of mixing scientifically determined quantities of sand, aggregate (stones), cement and water together.

Above: aerial view of a cement works. On the left are the quarries, from which raw materials are taken on a long conveyer to the plant. Centre right are groups of cement silos; foreground right is a small group of slurry tanks next to a long cylindrical kiln.
Below: a wet process kiln viewed from the firing end, where the dried slurry is burnt to make clinker.

Although it is widely used in the construction industry, cement only represents about 3% of total building costs.

Concrete

Concrete is the most widely used structural and civil engineering material today. Its applications range from small objects like fence posts and street light standards to roads, dams, cathedrals and massive offshore oil production platforms. The raw materials used for making concrete are found in abundance throughout the world and its technology is as suited to labour intensive and low-technology applications in the developing world as it is to the capital intensive and highly mechanized technology of the industrialized nations.

Basically, concrete is a 'conglomerate' of strong but chemically inert aggregates, that is, natural sand and small stones, or artificial mineral materials, bound together by a matrix of mineral cement. Cement hardens and gains strength over a period of time, as a result of chemical reactions with water, but before it hardens the ingredients for concrete can be mixed into a plastic mass and cast or moulded into virtually any shape.

The ancient Egyptians used hydrated lime and gypsum (calcium sulphate) cements. The Romans discovered that the addition of *pozzolana*, a natural volcanic ash found near Mount Vesuvius, produced a concrete that was not only stronger and more durable but would also set and harden under water, which made it valuable for building bridges and aqueducts. Lime and pozzolanic concretes continued to be used after the Romans by builders in the

81

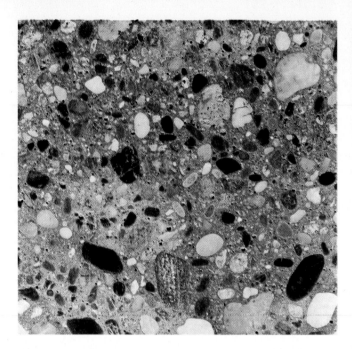

Concrete made with river gravel, as opposed to blasted gravel; note the smooth edges on the stones. Cement paste and water bind the particles together, and the consistency of the fresh concrete allows it to fit the framework closely and to compact easily.

A hardened cement paste has been reinforced with plastic fibre and the cement itself modified with polyvinyl acetate (PVA). This material has greater shock resistance, flexibility, tensile strength and breaking point than the conventional steel-reinforced matrix.

so-called Dark Ages (from about the fifth to the eleventh century AD). But it was the invention of Portland cement by Joseph Aspdin of Leeds, in 1824, that created new possibilities for concrete as a structural material.

Concrete is widely used in building for its inherent strength under compression. This is still the major controlling factor in modern concrete construction, and it depends mainly on two factors: the proportion of cement to aggregate in the mix, and the proportion of cement to water. In the broadest terms, the more cement in proportion to aggregates and the less water in proportion to the cement, the stronger the concrete. The proportioning of the mix, however, is intimately tied up with the application and the materials most economically available.

The standard method of testing concrete for compliance with its specification is the compressive strength or 'cube' test; a 4 or 6 inch (10 or 15 cm) cube cast and cured under rigid conditions is pressed to failure in a standardized testing press, most commonly at an age of 28 days. The material must be able to withstand pressures from 20 N/mm² (2900 psi) for lightly loaded floor slabs to 50 N/mm² (7250 psi) or more for ordinary structural concrete, and even higher pressures for some applications.

The proportions are generally chosen to give a combination of structural strength in the hardened concrete together with the necessary workability in the fresh, plastic material. An ordinary medium strength concrete for normal structural use might have an aggregate to cement ratio of between five and six to one, with a proportion of fine to coarse aggregate of between 3:7 and 4:6. A water content amounting to 25% by weight, or even less, of the cement content is sufficient for the hardening of the concrete, but additional water is needed for workability, and the water content of concrete is normally between 45 and 50% by weight.

Important to the strength, durability and appearance of the finished concrete is the care that goes into the mixing, placing and subsequent treatment of the mix while it is fresh and plastic. The various ingredients must be thoroughly mixed and for all but the very smallest quantities this means machine mixing. This is done either on the site or increasingly in central batching and mixing plants with the fresh concrete taken to the site in revolving drum ready-mix trucks.

Once delivered the concrete must be carefully placed in its formwork, so as to fill the forms properly and prevent separation of the ingredients. Traditional methods of placing by wheelbarrow or crane-handled skips are still widely used, but increasingly the fresh concrete is transported through pipelines and placed by specially designed pumps. In some applications, such as in swimming pool construction, it may be sprayed into place.

Thorough compaction is necessary to ensure complete filling of the formwork and, more importantly, to expel all the unwanted air from the mix. For relatively thin concrete slabs—building floors and road pavements—the concrete can be compacted by vibrating beams across the surface; internal 'poker' vibrators are used for larger and deeper sections. In both cases the vibration fluidizes the plastic concrete while consolidating it, allowing trapped air to rise to the surface.

Whether the concrete is finished smoothly by hand or mechanical trowelling or left with the texture imparted by the formwork, it requires careful curing. Contrary to what many people think, concrete does not harden by drying out; if it is allowed to dry too fast in its early life it is likely to be ruined. Preventing it from drying too fast, by slowing evaporation, is a key part of the curing process. This may be done by sprinkling the newly hardened concrete with water and covering it with polyethylene sheeting, or sometimes in the case of small objects such as blocks or pipes, high pressure steam curing in an autoclave is used.

Although concrete is very strong in compression,

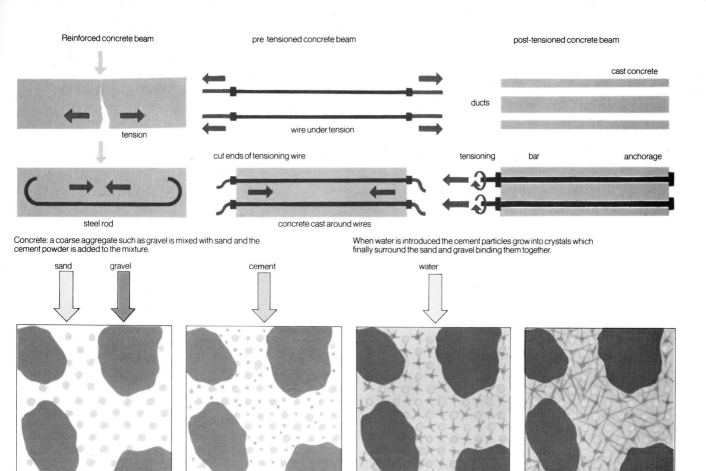

Reinforced concrete beam

pre tensioned concrete beam

post-tensioned concrete beam

tension

steel rod

wire under tension

cut ends of tensioning wire

concrete cast around wires

cast concrete

ducts

tensioning   bar   anchorage

Concrete: a coarse aggregate such as gravel is mixed with sand and the cement powder is added to the mixture.

When water is introduced the cement particles grow into crystals which finally surround the sand and gravel binding them together.

sand   gravel

cement

water

Above: these diagrams describe the mechanical properties of concrete as used in construction.

Below: concrete boats are popular in under-developed countries because they are cheap, durable and unskilled labour can be used in building them.

it has a limited strength in tension and bending. Until methods were developed for reinforcing concrete in tension, towards the end of the 19th century, the material was limited to foundations, footings, block masonry and other uses where the stress was almost entirely in compression.

Reinforcement provides the necessary tensile strength by embedding in the concrete another material which is strong in tension but, by itself, not well adapted to taking compressive loads. Most reinforcement consists of round mild steel bars positioned in the concrete.

The bond between the concrete and the reinforcement is very important and as a result square twisted bars and ridged 'deformed' bars are widely used to increase the bond. High tensile steels are also widely used where reinforcement must be especially strong. The design, fabrication and fixing of reinforcement require a high degree of technical skill, and it is fair to say that modern concrete depends nearly as much on steel—its supposed 'rival'—as it does on itself for the finished result.

Concrete does not need to be cast entirely on the construction site. One of its great advantages is that individual beams, planks, blocks or whole wall units can be cast away from the actual site and brought in as needed.

Besides reducing on-site work in congested locations, precast construction permits casting itself, finishing and detailing to be carried out under 'factory' conditions, affording standards of quality control that would be virtually impossible in on-site construction. Also standardized beams and other units can be formed in long life, high precision steel moulds for use in a wide variety of structural and architectural applications.

Of all the many techniques employed in concrete construction the one that can truly be said to be 'twentieth

83

Interior view of the Pantheon, Rome. The dome is made of light concrete; volcanic pumice was used in the mix. It was built between 120 and 124 AD.

Aerated concrete in a mould. Using a chemical reaction in an autoclave and high-pressure steam, a cellular structure of concrete which floats is produced.

century' is prestressing. Developed in principle and used in limited applications from World War I onward, prestressed concrete has only approached its full potential in the last 25 years.

Prestressing is a logical extension of reinforcement. The difference is that instead of merely passively resisting tensile and bending stresses as plain reinforcement does, prestressing counters them actively, allowing a lighter element to carry the same load. High-tensile steel bars, strands or cables are tensioned and anchored at the ends or at points along their length so that they keep the prestressed concrete element under constant compression within the limits of its design loading.

There are two basic forms of prestressing. In pretensioning, the concrete element is cast around tendons, prestressing wires which have already been placed under heavy tension between permanent anchors at each end of the casting bed. When the concrete has reached the desired strength and is bonded firmly to the tendons, the ends are cut off the prestressing wires and their elasticity places the whole unit under compression. Because of the way it is done, pre-tensioning is for practical purposes limited to precast units, but within those limitations it has various uses from fence posts and lintel beams a few feet long to bridge beams measuring up to about 150 ft (46 m) and weighing many tons.

The other form of prestressing—post-tensioning— involves the use of tendons or bars which are tensioned once the concrete has hardened by means of hydraulic jacks and specially designed anchorages. Post-tensioning

can be used with precast elements but its most widespread use is in on-location concrete structures. Preformed ducts are cast into the concrete during construction, the tendons are threaded through them and tensioned once the concrete has reached a predetermined strength.

Another use of post-tensioning is in segmental construction, especially suited to bridges and viaducts where site conditions prevent the use of complex falsework (temporary supports) between piers. The actual bridge deck structure is made up of 'slices', sometimes cast on site but more often nowadays cast and cured in advance, either in a yard alongside the contract or in a precasting factory farther away. As each segment is completed or positioned it is tied back to the structure by post-tensioned tendons; the whole structure is held together like a row of books picked up by pressing in on the end volumes.

Volume for volume, concrete is much lighter than most other structural materials, but strength for strength it is heavier. In many cases this does not matter, but where large spans are concerned, or in other uses where weight is a critical design factor, it can be a problem. The logical answer, and one the Romans discovered, was to use lightweight aggregates. For the dome of the Pantheon, slightly larger than that of St Paul's in London, they used a concrete with natural pumice aggregate, probably the first lightweight concrete in history.

Pumice aggregate is still used where it is easily available, mainly in lightweight blocks for house construction. But there are many other forms of lightweight concrete. Perhaps the most widely used is concrete made with

Reinforced concrete being used to lay the foundation of an office block in central London. Reinforced concrete is one of the most important of construction materials.

Reinforced concrete sections of a highway viaduct. The shuttering in the background is used for casting the concrete on site; it is moved each time and re-used.

lightweight aggregates such as clay, shale and slate, sintered pulverized fuel ash and foamed blast furnace slag.

Another form of relatively lightweight concrete is 'no-fines'—concrete made with normal dense aggregates but without fine material so that empty spaces are left between the coarse aggregate particles. Finally there is aerated concrete, actually a sand-cement mortar which is foamed chemically, either on site or during the manufacture of precast building units and blocks, to produce a very light cellular material which has reasonable structural strength, a high thermal insulation value and such a light weight that it will actually float in water.

All of these different forms of concrete have their specific applications. But besides these, there are special concretes with extra-hard aggregates for use under highly abrasive conditions, and other concretes with radiation-absorbing aggregates for use in nuclear power station.

Bitumen

Bitumen is a waterproof and adhesive substance that is relatively cheap, an advantage that gives it a wide variety of uses. Chemically, it is a complex mixture, essentially of hydrocarbons (compounds of carbon and hydrogen). It may be solid or a very viscous liquid, and it softens when heated. In its naturally occurring form, bitumen has been used for thousands of years as an adhesive. It is still principally used for road building, and also for waterproofing—an application that ranges from its use in roofing felts to the impervious layers in dams.

Bitumen was originally obtained from naturally occurring deposits, chiefly in the Middle East. Now most of the world's production of bitumen comes from oil refining. After the petrol [gasoline], paraffin [kerosene] and diesel fuel fractions have been distilled from the crude oil, suitable residues can be further distilled under vacuum to give bitumen.

The percentage of total bitumen production which is used for road building varies from country to country; in Britain it is about 70 %. Bitumen is not used on its own on roads: it is mixed with small stones, or with stones, sand and chalk filler. This mixture is called asphalt in Britain. It is confusing that in the United States the name asphalt is used interchangeably with bitumen to describe the bitumen itself, without any added material.

Asphalt (in the British sense of the word) was not the first way in which bitumen was used on roads. The name of the 19th century Scottish road engineer John Mac-Adam is perpetuated in the description *water macadam*. This was a mixture of graded small stones rolled in place on the road surface with sand and water, which helps to bind and compact the surface. Later, coal tar was used instead of the water, giving better binding and also waterproofing the surface (this kind of road was called *tar macadam* or 'tarmac' for short), and later still bitumen was used instead of the tar. Bitumen macadam is still used but asphalt is a more important road making material. The amount of bitumen in asphalt is small: between about 8% and 13% of it in the mix is enough to bind the components together and to make them waterproof.

There are other forms of bitumen mixtures used in road building. One of them is mastic asphalt, made up of sand

filler and bitumen. It gives a strong, smooth, exceptionally hard wearing surface, able to stand up to traffic for 50 years, but it is expensive and difficult to lay.

Bitumen may be used as a surface dressing on roads, to make repairs and, with small stones, to extend road life by renewing the wearing surface. It may then be used as a 'cut back' with a petroleum solvent or made into an emulsion with water. These methods allow the bitumen to be applied without heating, the solvent or the water later evaporating to leave the solid bitumen.

Roofing and floors
Roofing and floor applications use about 20% of total bitumen production. Roofing felt is the largest use; this is felt impregnated with bitumen and then coated on both sides with oxidized bitumen. To oxidize bitumen oxygen is blown through heated bitumen, making it harder and more resistant to weathering. The coated felt is dusted with talc to prevent sticking, and it may have sand or fine aggregate added to one side to improve weathering.

Mastic asphalt, mentioned earlier, can also be used for roofing. A roof at the Victoria and Albert Museum in London made with mastic asphalt is still in good condition after more than 70 years. Finished roofing felt may be cut up into square or oblong shingles; this is more usual in North America than in Britain.

A type of backed linoleum, and some kinds of tile, are made by impregnating felt with bitumen and applying a hard-wearing surface layer of resin with a decorated finish.

Other applications
Nearly all bitumen applications are, to some extent, waterproofing ones, but more specific waterproofing uses include an impervious screen in dams and preventing seepage from canal beds.

Above top: small granite chippings coated with bitumen are laid by a spreader on top of an asphalt surface. The chips give the road texture and skid-resistance.

Above: hot bitumen is used to bond a pitch polymer roofing material to an asbestos-based bituminous roofing.

Bitumen is used as an insulator and again as a waterproofing agent in electric cables and junction boxes. It is used in storage battery cases and seals. Anti-corrosive coatings on pipelines and other structures use bitumen. So do laminated wrapping papers, to which bitumen gives strength and waterproofing.

Bitumen has been used to stabilize desert sand dunes, allowing vegetation to become established, and as an agricultural mulch, absorbing heat and reducing evaporation from seedlings. It is used in black paints and even to seal coffins, a partial reversion to one of its first uses—preserving mummies.

Demolition

The art of demolition is almost as old as the art of building, and the safe demolition of buildings and other structures requires a great deal of skill and expertise. In general, the main ways of demolishing a structure are: dismantling it piece by piece, pulling it down or pushing it over, causing it to collapse under its own weight (often with the assistance of explosives), knocking it down with sledgehammers or a demolition ball, or various combinations of these methods.

Preparation

There are many factors to be considered before the method of demolishing a particular structure can be decided upon. These include its size and its type of construction, the proximity of any adjacent buildings, the time and costs involved, and any particular hazards present such as dangerous weakening of the structure by fire or flood damage.

Once the method has been decided, the first step is to arrange for the disconnection of electricity and telephone cables, gas and water supplies, and drainage and sewer connections. Road and rail bridges often carry cables and pipelines, and these must be disconnected and re-routed by the authorities concerned.

Saleable scrap such as electric cables and copper or lead pipes may be stripped from the building before the main demolition work begins, and many other re-usable items can often be reclaimed from a demolition site as work progresses.

Temporary supports, for example struts and braces, may be needed to prevent an uncontrolled collapse of the structure during demolition, and these, together with protective screens to prevent damage or injury by falling debris, are erected before the main work begins.

To help dispose of the debris from the upper levels of a building, disposal chutes may be provided, and where possible flooring may be taken up to leave a hole through which the debris can be dropped down to ground level.

Hand demolition

Demolition of buildings by the use of hand tools, including sledgehammers and pneumatic drills, is ideally carried out in the reverse order to that in which the building was built. Scaffolding platforms are erected, where practicable, to provide a safe area for the operatives to work from. Small structures, such as houses, may be demolished entirely by hand methods, but for taller structures, including chimneys and cooling towers, hand demolition

Dismantling the Temple Bar gate, London, in 1878. It was carefully taken down and later rebuilt on a country estate. Similarly, London Bridge was rebuilt in Arizona.

may only be used for the upper sections. This is continued until the height of the structure has been reduced sufficiently to allow the remaining part to be safely demolished by mechanical means (ball, pusher or wire rope).

The roofing is removed first, and heavy beams and girders are cut into sections that are lowered to the ground by a crane or hoist. Large slabs of reinforced concrete are cut into strips, with the cuts running parallel to the main reinforcement. Checks must be carried out before demolishing load-bearing walls or beams to ensure that their removal will not cause an uncontrolled collapse of the remaining structure. Uncontrolled collapse can also be caused by the pressure exerted on the walls by a build-up of fallen debris, so this must be regularly cleared away before a dangerous condition arises.

Mechanical demolition

A widely used method of mechanical demolition is by means of a demolition ball, a heavy, pear-shaped cast steel weight suspended from a crane or dragline excavator. The ball is swung against, or dropped on to, the structure to be demolished, but as it cannot be safely used on structures over about 100 ft (30 m) in height, the upper sections are first removed by hand demolition. As most cranes are not built to withstand severe shock loads their use in this type of work is limited to dropping the ball onto horizontal areas such as concrete slabs.

A pusher arm, as the name suggests, can be used to push over brick or masonry structures. The arm is fitted to a large hydraulic excavator in place of the bucket, and is typically about 15 ft (5 m) long. As the pusher arm must be applied not more than two feet (60 cm) below the top edge of the wall being demolished, hand demolition may be needed to lower the top of the wall to within reach of the pusher. The latest pushers have a reach of about 45 ft (14 m).

A structure or a part of it may often be demolished by attaching a wire rope or ropes to it, and pulling with a heavy or tracked vehicle or a securely fixed winch. A variety of structures may be demolished in this way, including brick and masonry walls and steel pylons and masts.

Deliberate collapse

In certain circumstances, where the site is level and well away from adjacent buildings and the structure is of suitable construction, demolition may be accomplished

Above left: dismantling a chimney. The rubble can be dropped into the chimney, but then it must be regularly cleared from the bottom or it will exert a dangerous pressure at the base of the structure.
Above: demolishing houses using a wrecking ball.

by the removal of key parts of the structure, causing it to collapse under its own weight. As with all demolition procedures, this is potentially very dangerous and must be carried out under expert supervision.

In the demolition of steel structures, a controlled collapse can be induced by weakening the supporting members and then using a pulling rope to start the collapse. One way of weakening the supports is the thermal reaction process. A mixture of a metal oxide and a reducing agent is packed around the girders or columns, and then ignited electrically. Large quantities of heat are generated by the reaction, which soften the steel to a plastic state so that only a small pulling force is needed to cause a collapse.

Another means of creating intense heat for cutting steel (and concrete) is by the use of a thermic lance. This is a long steel tube filled with steel rods; the end of the tube is preheated, and oxygen is passed through it, reacting with the steel at the heated end and creating an intensely hot flame. One drawback is that the lance burns away rapidly, so protective clothing and goggles must be worn by the operator to protect him as he moves towards the cutting area. Thermic lances, which are very effective cutting tools, are now becoming widely used in many areas of demolition work.

Explosives and bursters

One of the most effective means of causing the collapse of a structure is by the use of explosives. The charges are placed in carefully chosen positions at the base of the structure, so that when they are detonated the supporting members or walls are blasted away and it collapses inwards. Power station cooling towers are well suited to explosive demolition, as they are widest at the base, and the whole tower can be made to fall within the perimeter of the base. The charges are usually detonated electrically, but as there is a risk of the detonators being triggered off by stray electrical signals (from nearby radio transmitters or by electrical storms), it is often necessary to use plain detonators with safety fuses. Explosives are also used to help remove old concrete or other heavy foundations, and for felling old factory chimneys.

In cases where large blocks of concrete or masonry have to be broken up, but it is not possible to use explosives, an alternative method is the use of *bursters*. These are either gas expansion or hydraulic types. The gas expansion burster consists of a perforated steel cylinder filled with a chemical mixture that vaporizes and expands when ignited electrically. The burster is inserted into a hole drilled in the concrete, which is fractured by the considerable force exerted by the expanding gases when the burster is fired.

The hydraulic burster also fractures the concrete by expansion, but in this case the force is exerted by pistons placed radially around a steel cylinder and forced outwards by hydraulic pressure.

Post-tensioned concrete

Many modern buildings are supported on horizontal concrete beams that are internally tensioned by steel tendons running in ducts. The ducts curve downwards so that the centre of the tendons is lower than the anchorages at each end, forming an inverted bow shape. By applying tension to the tendons an upward force is created because of this curvature, and tension is progressively increased during construction of the building so that the downward load of the structure is balanced by the upward force in the tendons. This creates serious problems when the building has to be demolished: if the load on the beam is reduced the tension in the tendons can force the beam upwards and shatter it, causing an uncontrolled collapse of the building. If tension is removed from the beam it will no longer support the building and a collapse will again occur.

In order to prevent these dangerous situations arising, the tensions in the supporting beams must be gradually reduced to correspond with the reduction in loading caused by the demolition of the superstructure of the building, so that a state of equilibrium is always maintained. As there are several tendons in each beam, reduction of the tension in a beam can be achieved by successive cutting of the tendons.

Before demolition can commence on a building of this type, it is essential that the position and function of all the post-tensioned beams (so called because the tension is applied after the beam has been positioned) is known. It is vital that proper records are kept during construction so that they can be referred to when the building is demolished.

Demolition of all types is inherently dangerous, and must only be carried out by experienced workers under expert supervision.

Left: shoring supports a roadway after demolition of an adjacent building.

Below left: a wrecking ball in use. It can be swung against the structure or dropped on horizontal surfaces.

Below: an old church spire at Beckenham, England, being demolished using explosives. Church spires are hollow with a solid stonework peak.

Design

Centuries ago, when human beings built their own homes and made their own tools, they did these things the way their forefathers had done them, and over the centuries the design of everyday objects had evolved to the most useful and pleasant shapes. Objects such as furniture, firearms and musical instruments, for example, were sometimes designed and made with such care that they became works of art, simply because they were beautiful objects. Nowadays, however, the accelerating rate of development of technology and the increasing specialization of labour has led to a situation in which most of the things we use are designed and manufactured by somebody else.

Designers have a profound effect on the quality of life, and a correspondingly profound obligation to do their job well. Good design means that an object will be used longer, with more efficiency and with more pleasure.

Ergonomic design

Ergonomic design is the attempt to design consumer goods, industrial processes and architectural space so that these things can be used with a minimum of stress. Its attitude is that the human factor should be taken into consideration.

The development of the design philosophy now called ergonomics was slow during the period of world-wide economic depression during the 1930s, but received great impetus during World War II. To illustrate the impact of the war, it may be useful to compare the early motor car with the modern aircraft.

Early cars were noisy, unreliable and subjected the driver and his passengers to a great deal of vibration and exposure to the elements. They were designed at the convenience of the builder; sometimes the brake lever was actually located outside the driver's compartment, so that he had to reach out to set it. But early cars did not go very fast, and there were not very many of them; the owner of a car derived his pleasure partly from the ownership of a novelty.

By contrast, during World War II men were flying hundreds of miles an hour over long distances in circumstances of discomfort and extreme danger. Their seats had to be designed to provide as much comfort as possible in cramped circumstances; instruments had to be located so that they could be read at a glance; vital equipment such as oxygen masks in high-flying aircraft had to be well located and convenient to use. The human factor had to be taken into consideration, as well as cost, ease of manufacture, availability of raw materials, and other factors. In addition, ships and planes and other war material had to be produced in large numbers, using the most intense methods of mass production. During conditions of total war, the survival of a nation could depend on the design of its tools and the stress on its workmen and soldiers.

The word ergonomics was coined in 1949. It is derived from Greek roots, and literally means combining work with natural laws. Ergonomic design is also called *bio-*

Above, top: the conventional QWERTY typewriter keyboard was designed, consciously or unconsciously, so that the typist could not work faster than the machine could respond. It is claimed that if the DSK (Dvorak Simplified Keyboard, second picture) were adopted, typing speeds could go up by as much as 35%.
Above: a set of eating utensils designed for handicapped people. Access to public places - doorways, stairs and so forth - for the handicapped is another function of ergonomics.

Above: the Moulton was one of the first of the small-wheeled bikes, light but strong, with a low centre of gravity, and adjustable for any height.
Right: the controls on the tractor (top) are poorly designed; the machine in the second picture is much safer and more comfortable.

technology, and is a part of engineering psychology; while engineering psychology can be misused, as when a product is designed with unnecessary features in order to persuade people to buy it, the point of ergonomic design is that the aesthetic value of a well-designed product or process arises out of consideration for the people who have to use it.

The senses

There is evidence that a great deal of stress is caused by overloading of the senses. Hearing, vision, and the other senses are inputs into the nervous system, and if people are overloaded with sensory input they become irritable, frustrated and unable to concentrate.

Noise pollution is becoming more and more evident in today's world; many governments are passing laws to regulate the amount of noise which can be generated in a public place. Employers are required to provide protection in the form of ear plugs for people who have to work in noisy areas. A loud noise can damage an eardrum, but most hearing loss is caused by progressive deterioration of the inner ear, caused by constant exposure to noise: the part of the hearing that is damaged will be the frequency range of the offending sound. Many men have lost part of their hearing as a result of prolonged exposure to gunfire during military training.

The physiology of hearing has been intensively studied, but less is known about the psychological stress induced by noise pollution. The designers of architectural space and consumer goods have to be able to deal with it. Furniture, carpets and curtains can all absorb sound, and rooms can be built with walls that absorb nearly all of it. The complete absence of sound, however, is not desirable either; the ears apparently need some small amount of background sound to distract them.

From the point of view of the designer, sound control is often out of his hands. Pneumatic drills for breaking concrete can be built which are relatively quiet, but this makes them more expensive, and contractors and local governments are reluctant to buy them for that reason. Much of sound control requires common sense and maintenance, for example the exhaust systems of automobiles. There is much that the designer can do, however, to prevent the generation of unnecessary noise. A machine

A well-designed fork lift truck. The knobs have diagrams.

91

designer will avoid mounting an electric motor or a gearbox on a large, hollow part of a machine, because this will amplify the sound; alternatively, the hollow part of the machine can be stuffed with sound deadening material, such as foam blocks. Rubber insulators between machine components are also used.

The designers of audio equipment must try to provide as wide a range of usable frequency response as possible, without peaks of excessive distortion. Otherwise the listener's ears will get tired of making the adjustments, and he will turn off his radio or record player, possibly without realizing that he is a victim of *listening fatigue*.

Vision is an aspect of design where a great deal can be done. Lights in a room, which at first appear to be adequate, will result in discomfort if they are a bit too bright or not bright enough. If there is too much glare it will be made worse by reflection on machinery or table-tops. If the lights flicker, even if the flicker is not quite noticeable, the result will be psychological stress and inability to concentrate on the part of the individual. In some instances people have suffered severe depression from living in rooms with bare light bulbs.

Measuring the light, however, is not all there is to ergonomic lighting; for efficiency at work and at home, many factors should be considered. The available light in a room depends upon the size and shape of the room and the colour of the walls as well as the location of windows and lights. Glare can be overcome to a certain extent by providing dull finishes on painted surfaces. There must be adequate contrast; for example a machine operator or a student reading a book should have more light on his work than on the immediate surroundings, or he will be distracted. On the other hand, a lampshade which blocks out too much light will provide too great a contrast. Other aspects which must be taken into consideration when designing lighting for work or play are persistence

The diagrams on this page are based on Henry Dreyfuss' research for ergonomic design of kitchens, offices and so forth. The greater need for such design is fairly recent because of the faster pace of modern life.

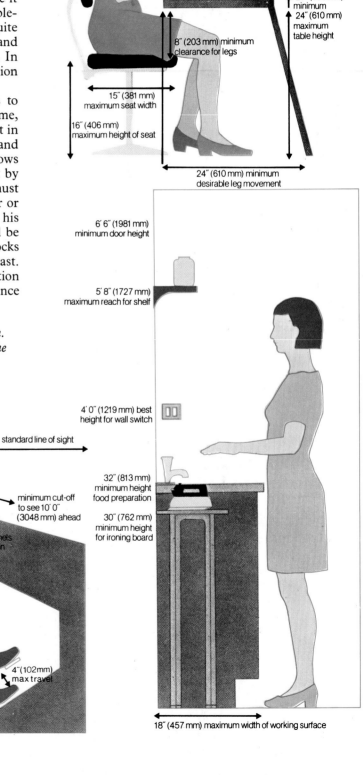

60° maximum easy head movement

125° maximum visual range without moving head

normal reach 23″ (584 mm) to 28″ (711 mm)

42″ (1067 mm) maximum see over height

27″ (686 mm) minimum
24″ (610 mm) maximum table height

8″ (203 mm) minimum clearance for legs

15″ (381 mm) maximum seat width

16″ (406 mm) maximum height of seat

24″ (610 mm) minimum desirable leg movement

6′ 6″ (1981 mm) minimum door height

5′ 8″ (1727 mm) maximum reach for shelf

4′ 0″ (1219 mm) best height for wall switch

32″ (813 mm) minimum height food preparation

30″ (762 mm) minimum height for ironing board

18″ (457 mm) maximum width of working surface

standard line of sight

minimum cut-off to see 10′ 0″ (3048 mm) ahead

control panels 90° to vision

15″ (381 mm) maximum height

4″ (102mm) max travel

17″ (432 mm) maximum seat width

of vision, location of objects which must be visually referred to, and the incidence of defective colour vision in the population.

Human skin and muscles are sensitive organs which quickly become aware of poorly designed furniture, inconvenient location of appliances, poorly balanced tools, uncomfortable temperature and humidity, and many other considerations. When a machine operator says, 'This feels just right', he means that his machine and its controls are well designed. Controls such as levers and buttons should be located so that they can be operated by the machinist without taking his eyes off his work. A control lever in the cab of a crane, for example, should move forward or sideways according to the direction in which the boom will move. The response of the controls should be positive without being too quick. In the kitchen, ergonomic design means that sinks and work surfaces will be conveniently located, and some thought given to the actual sequence of steps in preparing a meal. Work surfaces must not be too high or too low, to prevent backaches.

There can be nothing worse for a student or an office worker than an uncomfortable chair. Anatomical studies can be used in the design of good furniture. The seat and the back of a chair must be firm enough to provide support, and designed to provide it in the right places. The chair must not be too high or too low, or the individual will be forced to adopt a poor posture in order to reach the floor with his feet; this is why office chairs and piano stools are usually adjustable.

The senses of smell and taste are senses that most people can protect for themselves; nevertheless, designers today must resist the temptation to use chemicals to disguise unpleasant sensations or to provide imitation pleasant ones. A good example of the constructive use of the sense of smell is the addition of a peculiar smell to natural gas, which itself has no odour, so that leaks can easily be detected.

An example of the sort of pitfall a designer can fall into is the use of tinted glass in car windscreens. Car manufacturers will not paint their products dull colours, and sunlight reflected on the bonnet [hood] can be blinding. There is also the familiar problem of an approaching car with its bright lights on at night. But some of the various kinds of tinted or filtering glass which have been used to solve this problem have also cut down on night vision, a dangerous side effect. Solutions to design problems are not always as easy as they seem to be at first; the effects of the solutions must also be considered.

Architectural space

The subtle character of architectural space is one of the most important responsibilities of the architect. When he designs a building, he should be aware of all the subtle things that influence behaviour. When a person walks into a room, as for example the lobby of an office block, he is psychologically and emotionally affected by the size and shape of the room, the colour of the walls, the sound level in the room, the distance he has to travel to reach the receptionist's desk, and other factors. The company using the building can ask the architect to design it in such a way as to welcome visitors and make them feel as though they are invited inside. The type of fixtures in the room will determine whether or not visitors feel inclined to loiter there. If the room will be visited by a steady flow of traffic, as in a bank or a personnel department for example, the room and its fixtures can be designed to encourage the most convenient traffic flow. An expensive restaurant may want to encourage people to feel comfortable, while a lunch counter in a business district may want to discourage people from lingering over their coffee, so as to have the greatest possible turnover of trade during the peak hours of business.

Much of this is really engineering psychology. The ergonomic aspect is perhaps more immediately appreciable in the area of domestic architecture. If a flat is designed so that rooms are located adjacent to one long hallway, the people who use the rooms will spend a lot of time walking up and down that hallway, and people using various rooms will feel separated from one another, more so than if the flat were designed with all the rooms surrounding a central area. The environmental aspect of architecture has a profound effect on the sense of well-being, or lack of it. A direct correlation has been discovered, for example, between the number of flats opening onto a hallway in a public housing project and the rate of crime or vandalism in that project. If there are too many doors on a hallway, the lower income people who are housed in such buildings will feel depersonalized, as though they are living in a cell block rather than in housing designed for human beings, and they will behave accordingly, though they may not themselves perceive the reason for their feelings of frustration.

One well-known project in the United States was supposed to be the very latest thing in public housing. Much was made of the fact that it was constructed of materials that would need a minimum of maintenance. The residents of the buildings felt as though they had no control over their environment, and were surrounded by hard, unyielding surfaces, as though the character of the buildings was more important than the character of the people. The flats in the project were all designed so that mothers could not look out the window and keep an eye on their children at play, and so the open areas around the buildings were abandoned to vandals and muggers. There was no place in the architecture for residents to rub shoulders and chat with each other on a summer evening, as they had done in front of the tenement houses which the new project had replaced. Within a few months, the project had been reduced to a shambles by the frustrated residents, a profound example of what happens when the human factor is left out.

Design training

The people who do all this designing are divided into many specialties. The most commonly acknowledged among them are interior designers, product or industrial designers, architects, engineers, urban and regional planners, urban designers, fashion/textile designers, and graphic designers. These all further subdivide into fascinating sub-specialties. Inigo Jones, the British architect, began as a set-designer in the theatre, for instance. Today he might have been a special-effects technician for the cinema. There are also people who design printed circuits for the electronics industry, which must be assembled under magnification by people of unusual dexterity. There are designers of very specialized medical equipment, such as artificial heart by-pass mechanisms. There are designers whose works are produced in the billions such as Roy Kroc's masterpiece, the 'Big Mac' hamburger, or Monsieur Bic's ubiquitous ball-point pen. Sometimes only one item is produced as in *haute couture*, car customizing, space satellite design, or do-it-yourself projects around the house.

Most historians trace the beginnings of 'modern' design to the Bauhaus, a school at Dessau in what is now East Germany. The influence of the approach, teaching methods and philosophy of the school under its first director, Walter Gropius (an architect), belies its short life of about ten years before it was closed by Hitler. (Hitler believed that the style of Imperial Rome more adequately represented the glories of the Third Reich.) Of course, industrialization was underway long before the 1920s, but design was often astonishingly behind the pace of technology during the period before the Bauhaus.

A famous example is that of the handle of the everyday flat-iron. Traditionally this had been made of wood turned on a lathe, but when technology made possible a cheaper, more durable handle of injection-moulded bakelite, the manufacturers continued to produce the same form for many years from moulds based on the old handles, so that the shape continued to have more to do with lathes than with the anatomy of the human hand.

The aim of the Bauhaus was to bring design into step with modern technology. Unfortunately their enormous progress was often flawed by their desire to create a machine-like style rather than to see in the development of new machines and technology better ways of making things fit the hand, as it were. Some of their very elegant and machine-like tea services, for instance, were virtually useless for pouring tea. In dropping design history from their curriculum they may have thrown the baby out with the bath water, for while the period before was ridiculous in its excess of decorative detail there were also traditions of lovely and efficient design quite adaptable for production in an industrial economy. Indeed, the utilitarian beauty of traditional Japanese packaging and American Shaker furniture design are only recently widely admired.

The noble intentions of the Bauhaus failed most clearly in the field of architecture, where they achieved their machine-like appearance by carefully covering brickwork with stucco. While it cost a great deal of money in the old days to have intricate decorative carving done, it also costs a great deal of money to have masons lay brick or apply stucco to infinitely greater tolerances than are necessary for structural integrity. Today this trait is so ingrained in the construction industry that it is very hard

British Ford's Dunton drawing office. Design of a new car precedes production by several years.

to get a random pattern or an uneven surface from a skilled worker. This over-precision represents a high hidden cost in our efforts to produce simple, useful buildings.

On the other hand, the design of buildings that not only look machine-made, but actually *are*, is only now becoming feasible. Unfortunately the construction industry is still largely at the stage of the flat-iron handle; the research and development departments of large materials companies are still hard at work trying to produce a metal shingle that looks and feels just like cedar. Design is catching up, however, and a new wall component is now on the market of foam-injected monocoque steel construction, permanently coloured however you like, that offers the same insulating properties in a two inch thickness that you would get from several feet of solid brick.

Architecture is always a special problem in design by virtue of the size of its pieces. That is, it is easy to re-tool a manufacturing process to produce a different kind of hand-iron but very difficult to change over a very large component such as a part of a building. Moshe Safdie's famous 'Habitat' housing project in Montreal for the World's Fair of 1967, for instance, turned out to cost twice as much per dwelling unit because the government sponsors cut the size of the project in half. Since the concrete units were poured and steam-cured in an on-site purpose-built factory, a great many units had to be built to achieve what is called 'economies to scale', or in other words, to amortize the cost of building the local factory. Since only half as many units were built as were planned, each unit had to bear twice the amortization cost as it would have otherwise.

The design process

To understand the design process we have to understand both sides of how an object becomes reality. The process of design is discontinuous by virtue of stages, and normally involves relatively few people—the designers, their clients and consultants. Design processing is more immediate and will likely involve large numbers of people—the users.

The first stage of the design process consists of receiving the intentions of the client, whether it is a government, an entrepreneur, a school committee, a family wanting a house, a corporation or whatever. These intentions may be quite specific or quite vague. They may be naive or well-informed. In fact, the clients may not have a specific item in mind at all but only an idea of how it may behave, as in 'something that won't spoil in shipping' or 'something we can sell a lot of'. This is called a performance specification and it is up to the designer to come up with a product that matches the performance of the idea presented to him.

After being presented with the problem, the designer then goes through a process of programming, which consists of both tightening up and amplifying the client's requests. Through his experience in translating needs into objects, and knowing what is possible and what isn't, the designer may even be able to substantively change the client's view of what the problem is. In this phase all the relevant information is gathered, sifted, analysed and ranked into priorities. Priorities are essential. For example, if two of the client's requirements are to have the design by 1st October, and that the object must be smaller than a breadbox, it does no good to satisfy all the other requirements if the object comes off the drawing board on 5th November and happens to be the size of a car. For complicated projects such as the design of a new town, this programming process may go on for years and involve hundreds of people.

After the analysis phase comes the synthesis phase, the very heart of design activity. Here the designer must take all the little bits of information that have been collected—and it may approach millions of bits with the help of computers—and make something out of it that accounts for as many of these bits as possible. Professional design education is crucial here because the mental processes utilized are truly different from the way we've always been taught.

All our schooling up until design education has been linear, atomistic and analytical. That is, we learn to read or to solve problems a piece at a time and in a certain order. Most of our homework is geared to analysis. Writing a school paper is a process of breaking problems, events or issues into their constituent parts and attacking them a piece at a time. Of course when the philosopher Leibniz *discovered* the calculus he discovered it all-at-once—but you can't *teach* it that way. Design synthesis has that all-at-once quality to it, and is most often taught in a studio setting where students solve hundreds of design problems under the direction of studio masters.

Some people say that you can't teach design at all and that the studio is just a good place to teach yourself. Regardless, the studio emphasis is definitely on learning by doing. The design process then is a cycle of synthesis-analysis-synthesis, each time trying to positively account for more and more of the bits of programming information, until you decide you're finished or, more likely, run out of time or money. Then the designed item is fabricated, printed, built, cast, grown or otherwise produced. The designer has, of course, taken all the possible methods of production into consideration during the design process and may in fact closely supervise the actual construction.

Analysis

Now comes a period of testing known as post-construction analysis, where the item is somehow measured as accurately as possible to see if it performs as everyone expected it to. For a consumer good this may mean test-marketing to see if anyone will buy it. For a carpet it may mean running an abrasive wheel over it thousands of times to see how it wears. For electrical devices it may mean testing for shock-hazards. For drugs it means long-term testing on animals or volunteer humans. In architecture, unfortunately, this phase is almost never possible, due to the traditional way of budgeting projects and to the enormous complexity of desired or anticipated performance. Most often the architect simply turns to the next project without ever really knowing how the previous one worked out, except for perhaps an aesthetic response from a journalist-critic, often based on photographs rather than a long-term on-site investigation.

Design process is further complicated by the presence of the ego (pride) of the designer. He often suffers from what is called the 'anxiety of influence' and seeks originality for its own sake rather than for better design. Let's take, for example, the case of the omelette—an ancient and classic recipe for food made from eggs. There is only one way to properly make an omelette and if one follows the rules it is always delicious and satisfying. However, the anxiety of influence drives certain designers (chefs) to modifications, such as adding cheese, bits of bacon, herbs or whatever. These designers are equally looking for a better omelette and for recognition for their brilliant originality, hoping that one day the 'Clarke omelette' or 'Les oeufs à la Clarke' will be famous around the world. This seems wasteful, conceited and foolish as long as the original, classic recipe is better or just as good. Except that one day such a chef will bake it instead of frying it, coming up with a quiche or a soufflé—a development that comes to be widely admired and consumed and that cannot be compared to an omelette at all. The designer is often forgotten completely (unfortunately for his ego) and the new product—the soufflé—begins its own period of 'wasteful' and 'egocentric' development, often to cries of 'why don't you leave things the way they are?'

Often, however, self-conscious 'flair' in design leads nowhere, in which case it eventually dies out, much like the tail-fins on American cars in the late 1950s. Unfor-

Two views of Banquet House, Whitehall, London, by Inigo Jones. Although the design is unmistakably his, it is also perfectly representative of its time and place. Although it draws on classical sources, it was built (c.1620) before modern consciousness of style.

tunately, such experiments in durable goods such as cars and buildings have to be lived with for many years to amortize their cost, as there is too much invested in them to throw them away or tear them down until they wear out. Theoretically, their long-term presence should at least help us not to make the same mistakes again.

Related to problems of originality are the issues of 'style' and the underlying tension in the world of design between art and technics. 'Modern' designers, that is designers at least since the Bauhaus, universally imagine themselves to not be working in a particular style. If it was anything at all, the Bauhaus was a reaction against style itself. Yet the Bauhaus style is now instantly recognizable. Throughout history there had been a smooth progression from one style to another: Classic Greek, Roman, Romanesque, Gothic, Renaissance, Baroque, Rococo, and so on. During the Gothic period, for instance, no one would have thought of building any

Moshe Safdie's 'Habitat' modular housing project in Montreal. It was built for the World's Fair in 1967. It was a financial failure because of government interference in the economics of the design process, but remains a good example of mass-produced housing.

other way. But in the 19th century a knowledge explosion occurred through the widespread use of moveable type and gravure that put all the styles in the hands of any designer who could open a book. This further led to designers travelling around to look at the designs of the past—a professional tradition that continues to this day. Suddenly designers began using whatever style they deemed appropriate: Gothic for churches, Classic for banks, Renaissance for universities, and so forth. Short-lived revivals of certain styles followed each other higgledy-piggledy as designers became self-conscious about style in a way that would have been unthinkable during the dominance of the styles themselves.

The so-called battle of the styles is still partly with us today. While modern designers pretend to abhor such considerations, it is nevertheless true that external observers, professional critics and laymen alike, have no trouble at all classifying designed objects into particular schools of thought. The British architectural critic Charles Jencks is fond of making charts showing the precise placement of contemporary architects according to their philosophical affinities. Of course very few are producing Renaissance design these days, but the Modern Movement has splintered into at least fifteen different recognizable styles as such. One can always tell Swiss packaging graphics when one sees it, as well as Italian furniture design from the late 1960s, or Japanese Metabolist (organic) architecture from the same period.

Some kinds of design are more susceptible to 'style' than others because of their publicness (openness to appreciation) or looseness of programme. Design process is always closer to one or another of the two following paradigms. In the 'coin on the table' paradigm the coin is either on the table or off it—there's no in-between. Similarly, a design for a printed circuit or a heart by-pass mechanism either works or it doesn't, and no one cares how it looks or what colour it is. (Nevertheless there are examples of both in the design collection of the Museum of Modern Art in New York.) In contrast, with the 'sweeping' paradigm one can either do a thorough job, a sloppy job, an artful job (to music?) or a lack-lustre job of sweeping and still claim to have swept the floor in all cases. House design is like that and so is graphic design. In both cases there is room for lots of style, either because the programme is so vague or because the basic programme is so easily accomplished that the designer feels driven to make more of it than there really is. This is the tension between art and technics. Are heart-pumps all technics? Are graphics all art? And if houses are to be pursued as art, for whose benefit is it done and at whose expense? New York architect Peter Eisenman, currently famous for fifteen minutes, designs incredibly complicated and bizarre houses that have their roots equally in the European *De Stijl* (early modern) movement of the 1910s and in images of computer-generated geometric transformations (but laboriously hand-drawn). Yet he has clients willing and even eager to have him design houses where a column literally takes the place of a dinner guest at the table.

One way out of the tension between art and technics for the modern designer has been to approach the design process as if it were nothing more or less than problem-solving. A good example of how this works when it works well is the Vertebra chair, recently designed by Emilio Ambasz and Giancarlo Piretti and manufactured by Openark B.V. in Italy. The problem was to design office chairs that would be comfortable to sit in for long periods of time, and they approached it as a problem, not as an artistic exercise, by making thorough studies of blood circulation, body-shifting cycles and bone structure. The result is a chair that is very comfortable to sit in for long periods of time. There's a bit of style in it too as they have attempted to make it look smartly orthopaedic. But it does respond very well to the fact that neither Cratchit's accounting stool, nor the traditional soft and cushy leather club chair, nor the stunning chrome chairs of the Modern Movement were comfortable for any length of time.

There are two problems with the problem-solving approach to design however. One is the suspicion that it is used to somehow 'elevate' the design profession from the mire of art up into the technocratic stratosphere where the designs are above scrutiny. That is, if it's all no-nonsense problem-solving—and if it seems to solve the problem—then there's nothing to criticize. The other, related, problem is in projects where there might be emotional content for the users (which we will discuss shortly) such as in a house. Here, if everything is expressed as a 'problem' there will likely be a good deal of emotional material concomitant with what a house ought to be like that gets passed over simply because these issues do not firm up as 'problems'. Again, it's a responsibility-avoidance situation for the designer and an opportunity for intellectual fraud: designing what you please to satisfy your ego under the guise of solving a limited number of rather obvious problems.

The other side of the coin involves *design processing* and instead of involving the designer, now involves the users, consumers or inhabitants. First of all, here's a quote from Robert Sommer, the American psychologist: 'Most of us experience the environment just outside of the focus of awareness.'

What does he mean by that? He means that for the most part we do not use or experience designed objects with the same self-consciousness that they were designed with. We all perform functions through exercising choice but seldom know or care why we make the choices we do. We may think we do not like where we work because the work is dull or the boss is mean, when in fact it may well be the depressing rooms we work in. We may not like to eat at home because, in fact, we do not like our table or the lighting—and not even be aware of it—or at any rate be able to isolate our dissatisfactions (or satisfactions, for that matter). We may like a restaurant not for its food or service but because it feels the way we would like our home to be—but can't seem to imagine it really being that way. Yet we think we like going there because of their veal or something else incidental to the mark. In fact, we all stumble about like this, responding much more to advertising and whim than we do to self-conscious efforts to determine how best to enjoy our limited resources.

Studies have conclusively shown that the environment affects us very much, whether we like it or not or whether we know it or not. For instance, if you would like to be chosen foreman of a jury on which you are serving,

The office chairs on the next page are a spectacularly successful example of what the design process can achieve. Smart-looking, they are also comfortable to sit in for hours at a time. This is also a good example of ergonomic design. In this case, anatomical studies were taken into account; sometimes the design process must make use of computers.

sitting at either of the chairs marked 'X' will greatly enhance the likelihood of your being chosen by your peers.

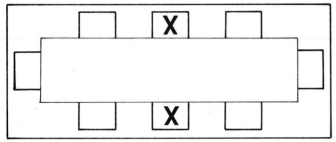

If the table is round, sit the farthest from the door to the room. This is an exercise in 'proxemics', or near-space, and has been proven again and again. Some people instinctively sense these things but most of us daily negotiate our environments with some clumsiness. It's very much like the old saying to the effect that we don't know who discovered water but we know it wasn't a fish. The issue at its largest scale has been beautifully expressed by the German philosopher Martin Heidegger who said, 'Before we build we must first learn how to dwell.'

Simply put, any effort we make to become more truly conscious of the effect on ourselves of the things or places that we use will be richly rewarded in the increased efficiency of our money, time and capacity for enjoyment.

Several things are at work here in our role as post-product designers (choosers) of the designed object. One is self-expression. If designers can express themselves through the design process, consumers can express themselves through choice as design processors. People tend to use objects in a way that best reflects their desire to express themselves. And it turns out, for example, that people who feel the need to design and build their own homes tend to be shy while people who are perfectly happy in furnished hotel rooms tend to be garrulous at the level of clothes and conversation.

The primary way in which our objects serve as vehicles for expression is through association. Associations between yourself and an object can be at any level of human experience: the abstract, the intellectual, the nostalgic, through kinship with nature, dreams, fantasy, gemütlich-keit (coziness) or whatever. Also, stronger associations are built simply by possession. A piece of wood-carving in a museum would mean a great deal more to you if you owned it and even more if you had carved it yourself. And all without any change at all in the object itself. Altering objects increases your associations too, which explains why people bolt on accessories to their cars (or completely rebuild them) or rearrange their furniture (or actually move walls or floors) or do such a simple thing as carve their initials in their school desk. It's as simple as people wanting their things to be like themselves (or how they would like themselves to be), as buffers and anchors in a hostile world, as a way of extending one's domain, as a way of knowing themselves. A cozy, warm sort of man may surround himself with pipes, fireplaces, big soft chairs and a stout old car with some personality to it. A technocratic mobile man-of-the-world would want slick, no-nonsense objects about with lots of buttons to push.

The American industrial designer Jay Doblin has taken the precepts of a school of psychology called transactional analysis and applied its theories to designed objects in a fascinating way. The theory holds that we are each made up of elements of the child, the adult and the parent in different proportions. By looking at a consumer market with a wide variety of choice to it (such as cars) Doblin has picked out three cars as best representing the constituent parts of the personality. They are, in respective order, the Mustang ('for the time of your life'), the Volkswagen Beetle (a 'practical' car), and the Cadillac ('for those in command'). When seen in the context of their advertising copy, which strongly reinforces the imagery, it is a very convincing presentation. All three cars can take you shopping. What kind of car do you want to have, and why?

It should be a humbling lesson to designers that their self-expression changes hands at the moment of creation when consumers take over to do precisely the same thing; but then they are already onto expressing themselves through their next design project. As soon as the product is produced, whether it's pantyhose packaging or an office tower, we the consumers could care less about the designer's ego investment. Their ego-massaging then happens among their peer group of other designers. What we *do* care about is our own ego investment—how well the product reflects what *we* care about—to the extent that we can know. Abstracted to the level of society it is trivially true that those products that are consumed the most mirror exactly that society's priorities. Studies to this point could be termed contemporary archaeology—and there are some such studies, but probably not enough. However, the process works only to the extent that we are truly aware of the effect on us of our objects and places—and of what they say about us to ourselves and to others.

The other necessary condition for this market to work is that there be a plurality of like-function objects for consumers to discriminate from. This again raises a special problem in the field of architecture. While we have an enormous variety of bars of soap to choose from—and even cars, due to economies of scale—we do not have that much choice regarding architecture, for several reasons. One is that the professional community is relatively small and therefore incestuous. While there are only 489 people for every doctor there are 3333 people for each architect (USA). This is rather a small cadre to whom to entrust the design of much of our visible, public, physical environment. Also, there are professional filters, such as standard national exams or curricula, and professional societies that give awards always to the same sort of buildings. No wonder much of it comes out looking alike.

It is an interesting note, now that we are entering the era of 'post-modern' design, as some call it, that several writers are speculating on the possibility that for the last forty years no one except the architects themselves and a very few like-minded critics have liked modern architecture at all. This raises the question, essentially an ethical one, regarding the apparently monolithic behaviour of the architecture profession: is it a good thing to keep trying for the quiche or the soufflé if the process in the meantime makes the omelette unavailable?

In short, there is less choice for the consumer/user in the field of architecture to enable him to express himself. In the special area of single-family housing (in which architects seldom participate in the USA) builders use the same designs over and over again because (1) a new design means slower progress by a work crew dealing with the unfamiliar and (2) most builders are small operators with a lot riding on each unit, which naturally tends to conservatism in design. So the market, within a given price range, has none of the variety found in soaps and cars. This reason, plus the price of new construction, is what drives young people on both sides of the Atlantic towards older houses, even if they are in need of substantial repair. There are more opportunities for self-expression in the range of older houses available.

In non-residential building architects do not deal with the users directly, but with the clients, who are more and more a faceless group of accountants, businessmen, lawyers and bankers. Clients who do not want to hear one dissenting voice from among their stockholders tend to insist on faceless buildings, like television producers who do not want to offend sponsors. Increasingly restrictive bureaucratically generated codes, standards and regulations also contribute, and so do problems in the work force. If you want to design a picturesque building, as we mentioned earlier, you will have a hard time finding a mason who can lay brick or apply stucco in an irregular manner and you likely cannot find a stone- or wood-carver at all. And if you want to design a very modern building with cost-saving unitized prefabricated bathrooms you will have to pay four plumbers to stand and watch the crane operator put it in place. These sorts of difficulties are not encountered by the graphic designer or even the appliance designer. They are quite free to innovate and often do. Furniture and appliance design are in fact miles ahead of architecture for these very reasons.

The last item to mention is something common to some extent to all professions. This is the concept of 'poets in the profession'. These are people in each field of design accounted as the 'architects' architect' or the 'engineers' engineer'. Oddly enough, these people are seldom the best of their peers in the strictly professional sense of the word. Their car designs may not allow enough room for luggage, their fashion designs may be terribly impractical as protection from the elements, their buildings may leak, their bridges may look dishearteningly fragile, their graphics may never cover a famous magazine. Often these people have trouble finding enough work to keep busy and are notoriously bad businessmen. But they serve a very important function by reminding the rest of the profession that design also has in it the power to lift the human spirit.

James Stirling's Cambridge University history faculty building. The use of glass is startling and the heating costs are high; it is a building about which one is forced to have an opinion.

Drainage

Drainage is the removal or control of surplus surface water, subsoil water or sewage. The main types of drainage are land drainage and town drainage, which includes sewage disposal.

In its natural state the earth can control its own drainage fairly well. Vegetational cover takes the initial impact from rain and protects the topsoil, whose humus (decomposed organic matter) content can absorb water and whose structure assists the percolation of the remaining water. The surplus flows into small streams which in turn produce a system of rivers, whose size steadily increases towards the sea.

Soils, however, vary greatly in the main single factor which affects their natural drainage, *permeability* (for example, gravelly soils may be 50,000 times more permeable than clay). (The permeability of soil is the rate at which water can seep through it.) In addition, higher lands usually have steeper slopes and more permeable soils than lower lands, and thus have better natural drainage. Man's usage of land, however, has tended to impair natural drainage. Soils comprise about half solid particles, and half pore space which is filled with air and water. Without drainage the pore space can become waterlogged. The uppermost surface of the waterlogged zone is called the *water table*; in good land the water table is deeper than 6 ft (1.8 m) from the soil surface.

The main objects of land drainage, land reclamation, erosion control, flood control, and prevention of waterlogging, are achieved by lowering the water table—that is, by removing the water from the soil or subsoil at a faster rate than it can accumulate.

There are indications that the Chinese used drainage systems for land restoration as long ago as 2300 BC. Since then localized independent efforts at drainage improvement have progressed to co-ordinated measures for control of surplus water throughout whole areas. In these land areas, known as catchment areas, all the water drains into reservoirs, lakes, or river systems.

Land drainage

The four main land-drainage methods are: open channel drains, underground pipes, 'mole' drainage, and pumping. In open channel drains or ditches gravity provides the force that removes excess water. They vary in size from small ditches, dug by hand or hydraulic excavators or trenchers and often seen around the perimeters of fields, to large channels 15 ft (4.6 m) or more deep, constructed by powered dragline excavators. The channels eventually feed into a river system, lake or reservoir.

Underground pipes were developed from 'covered tiles', first used in England in the late 18th century and made by bending an ordinary clay tile to an inverted 'U' cross-section before baking it and laying it on a flat tile to produce a channel. By the mid-19th century extruded cylindrical pipes were produced by machinery, and concrete pipes appeared a little later. The pipes, which are 2 ft (61 cm) in length, have spigot joints (one end of the pipe is larger than the other, and the small end of one

Above: deepening the sewer in Fleet Street, London, 1845.

Below: Mohenjo-Daro in the Indus valley had one of the most extensive drainage systems in history, 2000 BC.

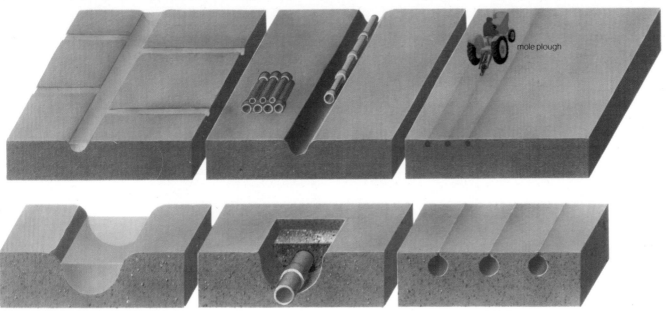

open channel drains   underground drains   'mole' drainage

pipe fits into the large end of the next) which allow some flexibility to cater for ground subsidence. The pipes are placed in temporary trenches about 2 ft (61 cm) wide by 3 to 6 ft (0.9 to 1.8 m) deep, generally made by trenching machines which are either self-powered or drawn by large tractors. Gravel or permeable filler is then placed on top of the pipes and finally the topsoil is replaced. Water flows into the pipes at the joints, which are protected by finely screened gravel, or tarred paper to resist any inflow of silt, which could cause a blockage. The pipes are usually arranged in parallel runs, known as main drains, fed at right angles by lateral runs. Spacing between the pipes is influenced by the permeability of the soil. The main drains are laid on a gradient and eventually feed through an *outfall* into a river system or reservoir.

Plastic pipes have recently been introduced, which can be wound into large coils and then laid by a suitably adapted trenching machine.

Mole drainage consists of moulding drainage channels into the subsoil without any artificial lining. The system requires a stone-free soil and an even slope. Channels, parallel to the ground surface, about 2 ft (61 cm) deep, 6 inch (15 cm) in diameter and 3 to 5 yds (2.7 to 4.6 m) apart, are made by a mole plough. This has a vertical knife blade carrying a horizontal bullet-shaped bottom member, was invented in about 1800 and is the forerunner of a machine now capable of fabricating concrete drain pipes in the mole channel, thus avoiding the need for digging and refilling trenches. The modern practice is to use mole channels in combination with pipe systems, the channels running at right angles to the pipes and feeding into the permeable material on top of the pipes.

Pumping water from below the water table is carried out in areas where the water level is so low, perhaps below sea level, that gravity drainage is not possible. The most common arrangement is for a pumping station to receive water from a low lying area and pump it to an embanked river and hence to the sea.

Town drainage

In towns, drains are needed for both sewage and for the surface water that is collected from the many impermeable surfaces such as roads and roofs.

There are three principal methods of town drainage.

Above: land drainage. Some of the earliest systems consisted simply of open channels or ditches forming networks across the fields, finally converging on a main ditch or river. Clay tiles or jointed pipes are used for underground drainage; water enters them from the surrounding gravel at the joins and flows away.

Below: this trench-digging machine for laying plastic pipe can also be hand-fed with unjoined clay pipe.

In the 'combined system' a single main sewer under the street carries all the sewage and all the surplus surface water to a sewage disposal unit. With this system it would be expensive to have main sewers large enough to carry all the rainfall which might occur in a heavy storm, and it would tax sewage disposal units to handle such large volumes. Therefore it is usual to provide storm water

103

overflows which discharge directly into rivers when the water level in the main sewer is too high. As the amount of sewage relative to the amount of rainwater is very low in these circumstances, pollution is not usually a problem, and the large volumes of rainwater have a cleansing action on the system. This system is simple and cheap, and is often used in rural areas where the amount of sewage is not high, and in coastal towns where untreated sewage is piped out to sea to be carried away by tidal action.

An alternative drainage method is the 'separate system' in which a main sewer under the street carries all the domestic sewage and industrial effluents to a sewage disposal unit, and a main drain carries all the surplus surface water into a river, lake or reservoir. This method has largely been superseded by the 'partially separate' system in which a main sewer under the street carries all the drainage from buildings (both sewage and the surface water from roofs) and a main drain carries all the surface water from the roads. Whichever system is used, rainwater from roofs is collected in a similar way. Rainwater is drained into a gutter around the roof perimeter by the slope of the roof, and is discharged through vertical downpipes into the drain system. Roads are built with a camber to drain surface water into roadside gutters. The surface water is then discharged at intervals through gullies into the drain system. On rural roads surface water is often led straight into open roadside ditches, or into 'French drains', which are trenches filled with pebbles or large stones through which the water can easily percolate. Embankments are often drained by French drains.

Drains vary in diameter from 4 inch (10 cm) for household purposes to several yards for main sewers, according to the amount of flow they are expected to carry. Small drains are usually constructed from iron or stoneware pipe and as size increases precast concrete is used. Trenching methods similar to those for land drainage are used but concrete foundations are usually provided to resist damage to the pipes by ground settlement. Also the pipe joints are made watertight by sealing the spigot with cement or bituminous compounds. If drains or sewers are particularly large or deep, tunnelling methods may be used. Drains are laid on a gradient to provide the required flow to the point of discharge. For maintenance purposes covered inspection chambers known as 'manholes' are fitted at regular intervals, providing access from the road surface into the drain.

Above: the entrance to a twin-tunnel storm water drainage system in Bristol, England, part of a scheme to prevent flooding in the low areas of the city south of the River Avon.

Below: paving the floor of a drainage tunnel.

Below left: a machine for cutting storm water tunnels. The rotating head cuts into the tunnel face, and the material removed is carried away on a conveyer to a rail car. The self-propelled machine runs on crawler tracks.

Flood control

In 1954, a hurricane in the Atlantic generated a tidal surge that struck the New England coast and resulted in the loss of 60 lives and damage of approximately $600 million. The following year the four major river basins of southern New England experienced devastating floods that killed 90 people and produced damage of $530 million. These are typical examples of the havoc that floods, whether from sea or river, can create, and are a justification for the massive investment in flood control projects.

It has been estimated that 12 % of the population of the United States lives on land that is subject to periodic flooding, and in Canada the proportion is similar. The amount of housing and industrial development on flood plains of rivers is still growing, so that damage and loss of life from flooding will increase unless measures to control floods are adopted.

River flooding

Natural river channels are formed by water (usually in the form of rain) draining from the land (this is called *run-off*). This water moves under gravity to the lowest point of the land and ultimately finds its way to the sea. At the low point of the land where the run-off is concentrated, a channel will be a function of the area from which the run-off is derived and of the rainfall that occurs. It will not grow in size indefinitely, but will, after a long period of time, reach a limit and not enlarge itself further.

The discharge (that is the volumetric rate of flow measured for example in cubic metres per second) that the channel can accept without flooding the adjoining land (called the *bankfull discharge*), occurs more frequently than once a year. The exact frequency is not known, but analysis of the discharge records of rivers in England and Wales has shown that the bankfull discharge is equalled or exceeded once every six months. It is recognized, however, that the bankfull capacity of a channel is less than the maximum discharge that the channel will have to deal with. It is inevitable then that flooding will occur.

When rain sufficient to cause a flood starts to fall on the catchment area of a river, the discharge in the river progressively increases with time, as the run-off from the more distant parts of the catchment arrives downstream. Because a rainstorm lasts for a limited time, the discharge in the river will gradually rise to a peak and it will start to decrease when the run-off from the catchment starts to diminish. This variation in river discharge with time is called a *flood hydrograph*. The flood can be regarded as a large wave which, as it travels down the river, changes shape, mainly as a result of the volume of water that is stored in the river channel and on the flood plain at any particular moment of time. The peak of the wave becomes flatter and the maximum discharge decreases. This feature is known as *attenuation* (reduction). It is possible to calculate how the flood wave will attenuate as it travels down the valley (routing the flood) although the computations are complex.

Protection measures against river flooding can be sub-

These two pictures show a newly constructed dam in the Netherlands. Above, earth is laid to form a bank along the proposed dyke. Below, woven mats are floated to the site and large stone finishing blocks laid on top.

divided into those that lower the flood water levels by reducing the peak flow of the hydrograph, and those that confine the flood to specific and well defined areas.

Peak flow reduction

One method of reducing the flood discharge is to reduce the rate at which water runs off the contributing catchment area: this can be achieved by land treatment measures such as afforestation, control of soil erosion, and improving agricultural techniques. These measures are intended to delay the direct overland run-off to the river and to increase the amount of moisture stored in the soil.

Reservoirs and flood confinement

Storage can also be provided by constructing a dam across the river to form a reservoir. The dam incorporates an overflow structure—a *spillway*—often fitted with movable gates, which is used to control the discharge that is released from the reservoir to the river downstream. At the beginning of the flood season the water level in the reservoir is at a level that provides sufficient room to store

floods without overtopping the dam. During a flood, water is released from the reservoir through the spillway at such a rate that serious flooding downstream from the dam is prevented: this release rate will be less than the rate at which water is entering the reservoir and the difference between the two flow rates will be taken into reservoir storage.

For economic reasons it is now usual to build reservoirs that fulfil a number of different functions such as flood control, irrigation, hydro-electric power and water supply, and chains of such multi-purpose reservoirs are to be found on some of the world's major rivers and their tributaries. Because of the multi-purpose nature of these reservoirs, it is very important that they are operated in the most efficient way, with the aim of maintaining the water levels in the reservoirs as high as possible but not so high that they will not provide a sufficient volume of flood storage.

In order to operate a chain of reservoirs as an efficient unit, it is necessary to be able to predict when a flood is likely to occur, what its hydrograph will be, and how it will attenuate in its progress along the river to the reservoir, so that the inflow into the reservoir can be determined. Also it must be known how the water released

Left: the Norris dam was the first built by the TVA (Tennessee Valley Authority).
Below left: a chain of reservoirs and dams.
Below centre and right: the planned protection from flooding for the city of London would include rising gates across the Thames and control channels.

chain of reservoirs

reservoir

dam

plains

lower cultivated plains

from the reservoir (which will also act as a flood wave) will attenuate in its passage downstream. This information is required quickly to allow sufficient time for releasing water from a reservoir before the flood arrives there, should this be necessary. Until fairly recently this was an impossible task because of the time required for the calculations, but by using computers it is now possible to plan reservoir operations more efficiently. Development work has been going on to improve the mathematical techniques that are involved, and some computer models of such reservoir systems have been built.

Two types of computer model can be used for studies such as this: a digital model which uses arithmetical operations to solve the equations that are involved, and an analog model which uses physical analogies for the variables in the problem, for example, an electrical analogy in which discharge is represented by an electric current, water level by a voltage or potential, and storage by the charge on a capacitor. For certain problems analog models are much faster than digital models.

One of the earliest computer models of such a reservoir system was that of the Kansas River, built in the early 1960s by the US Corps of Engineers. Digital and analog computer models were both built. Rainfall forecasts were fed to the computers which then determined the most effective way to operate the reservoir system, for example, the sequence of water lowering operations in the various reservoirs. Flood warnings were also given by the models.

Other methods
In the middle and lower reaches of a river there is often a lack of suitable sites for large storage reservoirs, and other methods of preventing flooding have to be used.

The channel can be enlarged so that its bankfull capacity is increased. This can sometimes lead to problems with sediment deposition, and often the amount of enlargement that is practicable does not give any significant lowering of the water levels for the more infrequently occurring floods.

Another method is to provide an additional channel at places where it is necessary to lower the water levels, for example through towns. The additional channel is so designed that it only operates when floods occur—some of the flood water spills into it, the discharge in the main channel is reduced, and hence flood water levels are lowered. The flood relief channel either rejoins the main river further downstream, or is linked to another river.

Finally it is possible to build levees—banks on the flood plain, which run alongside and close to the river channel. When floods do occur they are prevented from spreading over the flood plain.

At the present time there is insufficient knowledge to enable computer models to be built that will predict accurately the effect of these various flood alleviation measures. Instead, small scale models of the rivers are built, which accurately reproduce the behaviour of the river in nature and enable the most satisfactory way of providing flood protection to be determined. The Hydraulics Research Station at Wallingford in Britain has studied many such problems by means of small scale models. A model of the River Trent in Britain was used to determine the height and position of floodbanks that would prevent flooding.

overflow channels in towns

river barrier to prevent tidal flooding

raised banks

estuary

flood channel

rising sector gate

Tidal flooding

The effects of tidal flooding are similar to those of river flooding although the causes are different. Atmospheric depressions (low pressure zones) over the sea can cause the water level beneath them to rise significantly; in addition they will also produce strong winds. As the depression moves or weakens, it and its associated winds may cause a surge to form, which could create unusually high water levels along an adjoining coast if it coincided with a high tide.

A catastrophic example of this occurred in 1953 when a depression in the North Sea caused a surge to move southwards, producing a level in the Thames Estuary that was 2 m higher than predicted and a level at the Dutch coast that was also much higher than predicted. The surge gave rise to disastrous flooding in SE England and in the Netherlands and altogether over 2100 lives were lost. Protection against such tidal flooding can be provided either by the construction of sea defence walls that are high enough to prevent overtopping or, in the case of an estuary, by means of a barrier, usually in the form of movable gates that can be raised and lowered.

As a result of the 1953 disaster, the Netherlands government pressed ahead with their Delta plan, which is a scheme involving both fixed and movable barriers across the mouths of the Maas and Scheldt rivers: it was designed to protect the Dutch coast against a storm flood level that would occur only once every ten thousand years, and a very large part of the engineering work has now been completed.

In Britain, the Government decided to construct a tidal barrier in the Thames estuary in order to protect London against tidal surges. This barrier has been the subject of a number of different model studies. A small scale model and a digital model have been constructed at the Hydraulics Research Station and have been used to determine the best position in the estuary for the barrier.

Studies have also been made of the effect that different operating procedures would have on tide levels along the estuary, and of the effect that the barrier would have on tidal flow, salt water intrusion and movement of sediment. A digital model of the North Sea has been built by the Institute of Coastal Oceanography and Tides in Britain, for use in the Thames barrier investigations. This will use weather forecasts to predict surge heights in the North Sea and the resulting levels in the Thames estuary. When the barrier is built, this model will be used to plan the operating procedure in order to prevent dangerous flood levels from being generated along the estuary.

Above right: during the floods of 1963, workers from the oil refinery on the Isle of Grain, Kent, England, helped to breach the gaps with sandbags against future high tides.

Centre: a tide prediction machine at the Liverpool tidal institute. This was one of the first machines in the world to be able to predict the time and height of tides years in advance.

Right: using a scale model, scientists at the Hydraulics Research Station at Wallingford, Berkshire, have been studying the possibility of Thames flooding and how to deal

with it. Computers will also be useful in this kind of work.

Frame construction

Among the various forms of construction those which are based on a load-bearing skeleton or frame are of importance. These frame structures consist essentially of relatively long thin members, interconnected at joints to form an arrangement capable of carrying loads without undue distortion of shape. The way in which the members are arranged and the nature of the joints is important. If the joints are free to rotate (pin joints) then the shape of the frame can only be maintained by the correct number and assemblage of the members. If there are too few members the arrangement is called a *mechanism*, structurally unacceptable although mechanically useful. A sufficient number of members produces a simply stiff structure; any additional members are redundant (though there may be other reasons for including them) and the structure becomes over-stiff, being commonly called a *redundant frame*. In contrast a rigid frame retains its shape by the use of stiff joints which do not allow the members to move relative to one another.

When a frame is clad with some form of skin the existence of the frame may not always be apparent. The external appearance of both ships and aircraft, for example, quite belies the fact that under their smooth surfaces is a complicated frame. In this type of structure the skin may be no more than a light covering; it can however contribute to the strength of the frame, in which case it is termed a *stressed* skin.

The engineering attraction of frame construction is the very economical use it makes of material. In a massive structure much of the material is used uneconomically because it is not loaded to its full capacity. In an ideal framework, on the other hand, the members are so arranged that they all work to their full strength. It is possible in fact to arrange the members in such a way that the minimum weight of material is used. But although this may sound economic, in practice the com-

A stressed skin adds even more strength to this airframe.

plexity of the resulting structure, leading to complex jointing problems, is not attractive.

Frames are classified not only in accordance with the rigidity or flexibility of their joints but also in relation to whether they are sensibly two- or three-dimensional in construction. The three-dimensional arrangement is often termed a *space frame*.

The idea of frame construction is clearly very old; certainly early man used a form of tent in which a light skin covering served to keep out the elements.

The skin, carried on a framework of wooden poles, was purely an environmental shield having strength in tension but only able to retain its shape because of the stiff frame under it without which it would crumple.

Despite the simplicity of many frames, a considerable amount of design effort is required to produce an efficient jointing system. Poorly designed joints are a source of weakness; in addition the designer needs to pay attention to the effect of the joint on the appearance of the frame. The joint problem is at its most difficult in space frames in which a number of members may intersect at awkward angles and much ingenuity has been shown in devising standard joints for such situations.

The frame members may be formed from a variety of materials; steel, aluminium, timber or even reinforced concrete are used with a preponderance of steel frames. Metal members can be either hollow (round or rectangular tubes) or of solid rolled or built-up section. For ease of jointing, timber frame members are usually of rectangular section connected by bolts or nails.

Ships

A very early example of frame construction in boat building is to be found in the coracle, the existence of which has been recorded in every continent except Australia and Africa since early man, and which is still used in parts of Britain. Animal skin stretched over a slim wooden frame provides an easily assembled, portable boat. The modern sporting canoe is often of similar construction.

The modern ship owes much to the ideas embodied in the coracle, though steel plating, contributing not only water-tightness but also strength to its supporting frames, has been substituted for the skin. A typical ship structure consists of a number of rigid welded steel frames spaced some 60 cm (24 inches) apart, forming the shape of the hull and providing a skeleton to which the plating is attached.

Above left: cranes and ships' hulls are both examples of frame construction. The ships' hulls derive strength from the frame rather than the outer steel plating.
Above top: a space frame constructed on the Emmerich system. Such a frame derives its strength from its configuration; the individual elements can be thin.
Above: wooden frame houses are common in North America and Australia. This Australian house is being built on stilts. The horizontal members are usually thicker than the uprights.
Opposite page, top: the stressed skin principle. A thin horizontal shape will bend under a load, but if the load is applied to the edges it will stiffen. The Pyradex roof system uses units of stressed skin.
Opposite page, below: geodesic domes are constructed from a network of plane triangles built up to form a hemisphere. In double layer grids, diagonal struts connect members of lattice girders.

Aircraft and cars

As with ships, but over a period of seventy years rather than several thousand, aircraft structures have progressed from wooden frameworks covered with fabric to present day all metal construction in which the external envelope participates with its supporting frame to form a stressed-skin construction.

In the aircraft structure, where minimum weight combined with high strength is essential, efficient frame arrangements have been investigated in detail. Light aircraft are still built with a fabric covering over an

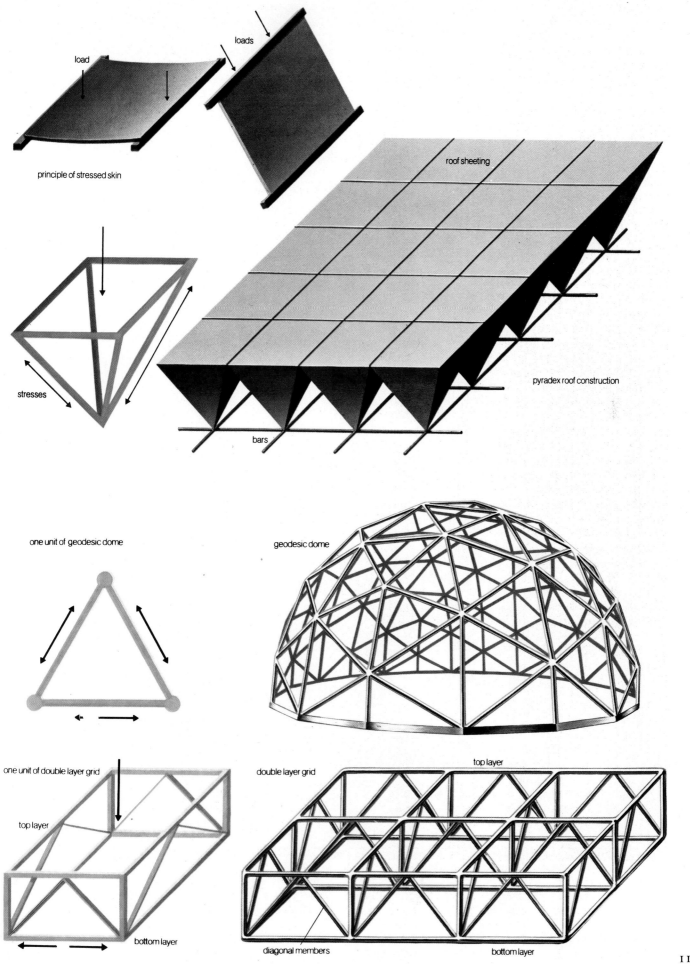

principle of stressed skin

load

loads

roof sheeting

stresses

pyradex roof construction

bars

one unit of geodesic dome

geodesic dome

one unit of double layer grid

top layer

bottom layer

double layer grid

top layer

diagonal members

bottom layer

aluminium alloy frame but the heavier types are more nearly surface or shell structures in which the strength is contributed largely by the stressed skin stiffened by ribs.

The structure of motor cars has also tended in recent years to discard the chassis frame in favour of a steel shell. *Monocoque* structures of this type are sometimes also used in racing cars, but many designs have now reverted to a space frame construction.

Buildings and civil engineering

On a domestic scale frame construction has a long history dating back to the charcoal burner's hut and the timber framed house. But in scope, and in terms of size and span it was severely limited until structural iron and later steel became available. The high strength to weight ratio of steel, in particular, has made it possible to build higher and span farther than ever before and frame construction with its simplicity and lightness has contributed to this progress. Thus the construction of buildings as high as the Empire State (1472 ft or 449 m) or bridges as long as the Quebec Railway Bridge spanning 1800 ft (549 m) have become possible. The major attraction of frame construction in such circumstances is that individual members are light, may be fabricated under cover in a factory and are easily connected together on site by bolting, riveting or welding. Frame construction is ideally suited to prefabrication. Today one sees frames in almost all aspects of building and civil engineering construction; timber roof trusses for houses, crane jibs, electricity transmission pylons, railway signal gantries, space frame roofs, and factory and warehouse building frames.

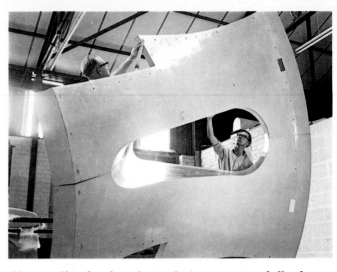

Above: a fibreglass form for producing a concrete shell to be used as an alternative to a frame. Concrete is an acceptable substitute in certain circumstances because it is very strong in compression.

Below: two views of the space frame roof of the new Covent Garden market at Nine Elms, London. Such roofs are both light and strong and can therefore cover large areas with minimum support. In the close-up picture, the members are thin but they are each taking only one-quarter of the thrust taken by the supporting column.

Land reclamation

Land reclamation in various forms has been carried out for at least the last five thousand years. Around 3100 BC the founder of the first dynasty in ancient Egypt, Menes, is reputed to have had a dam of stone blocks built across the Nile in order to build the city of Memphis on the drained area, with the diverted river forming a perimeter defence.

The farm areas along the Nile were divided into large flat basins by embankments of mud and stone several feet high. Canals and sluices admitted water from the river, which not only irrigated the ground but also deposited silt, thus improving the quality of the land. Other ancient civilizations depended for their existence and advancement on the regulation of rivers, notably the Babylonians in the Tigris-Euphrates valley and the Indians in the Indus valley. The Romans introduced reclamation to Britain, constructing earth embankments and draining the enclosed areas for agriculture, especially in the east of England around the Wash.

The reclamation of land from below sea level, to create *polders*, was achieved in Holland in the second half of the fifteenth century following the development of the windmill. After the construction of an earth dam, the enclosed silt or peat was drained by pumping the water into neighbouring land or into a drainage ditch. Since the windmills could raise water only about 5 feet (1.5 m),

double and triple lift systems were developed, the water being pumped into a storage basin between each windmill.

The use of windmills resulted in frequent flooding at times when there was little or no wind, but this problem was overcome in the late eighteenth century by the introduction of steam driven pumps, which were eventually replaced by diesel or petrol [gasoline] engined pumps. Today automatically controlled electric pumps are used.

A modern reclamation scheme can be an extremely large and expensive project. For example, the current Delta Project, to the south of the Hook of Holland, will shorten 450 miles (724 km) of coastline to 15 miles (24 km), with dams up to 5 miles (8 km) long across estuaries as deep as 140 feet (43 m).

Techniques

The first step in reclaiming land from the sea, river estuaries or other submerged areas is to construct a bund, or dam, around the area to be reclaimed. This may be done in a number of ways, depending on the size and location of the site and the availability of suitable materials. The simplest method is to use clay or sand bags or sand, possibly lined with clay or polythene sheeting. The sand is prevented from being washed or blown away by the use of grass or fibre matting or steel mesh.

An alternative method is to use gabions, first used in

Above: work in progress on a land reclamation scheme at Bijlmermeer in Holland. The level of the land is being raised with sand dredged from the North Sea.

Left: a map of the Delta Project in Zeeland, Holland, showing the areas to be enclosed. It is intended to avoid a repetition of the disastrous floods of February 1953 in which 1835 people were drowned, and requires control of the Rhine and Schelde estuaries.

the sixteenth century in the form of wicker baskets filled with soil. Modern gabions are formed from large plastic coated steel wire mesh boxes, usually rectangular in shape and filled with stones. Large schemes, particularly industrial projects, may involve the use of interlocking sections of steel, concrete or timber called sheet piles which are driven into the ground, large concrete boxes called caissons which are sunk into the ground, or a large deeply embedded concrete structure reinforced with steel bars which is constructed in situ.

Once the dam has been formed the area confined is ready for reclamation. The simplest method is to pump out the water and install a drainage system. Having done this, however, the risk of flooding remains, so it may be decided to raise the area above the level of the adjacent sea, river, swamp or marsh. This can be done by using a suction dredger to pump in a slurry of sand and water, called a hydraulic fill, from outside the bund. The water is continuously drained off, and the level of the reclaimed land gradually rises. Another method is to fill in the reclamation with stones, rocks, hardcore or any readily available local material.

After the area has been filled in the ground will still be soft, but in time it will subside and become firm. If the land is required for building purposes it may be desirable to speed up this natural process. This was traditionally done, as in the case of Venice, by placing heavy layers of timber on the soil.

Stabilization is now achieved by a variety of techniques, including compaction with a vibrating roller, injecting cement dust into the ground, driving columns of rock into the ground (vibroflotation), or the sinking of concrete piles into the ground. Surcharging is a method of compacting the ground by the use of sand, soil or water contained in polythene sheeting which is placed on top of the area and removed at a later date, possibly six months later, the weight of the surcharge forcing the water out of the soil and increasing its density. Another method uses fibre filter sheeting laid on the ground. The water is forced up through the sheeting, which allows water to pass through but holds the soil in place.

Desert reclamation

Arid areas, where the annual rainfall is less than 20 inches (50.8 cm), can be farmed only at risk of crop failure unless irrigation water is provided. The soils in the drier lands are relatively unleached because of the low rainfall,

Above: this series shows reclamation for an oil rig service base at Peterhead Bay, Scotland. Left: dredged sand has been pumped into the bay behind the breakwater by the dredger in the picture. Centre: a few months later foundations are being built. Right: work is almost complete and the quay is already in use.
Right: placing the final caisson to close off the Lauwerszee, Holland.
Below: a working island in the Easter Scheldt, part of the Delta Project.

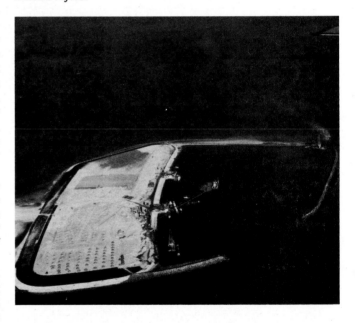

that is they are relatively rich in mineral salts as there is not enough water to wash them away. Where water can be made available in sufficient quantities for washing out excessive accumulations of salts, the soils in dry arid land can become highly productive, as in the Imperial Valley of southern California where 500,000 acres (202,343 hectares) were brought into cultivation.

In desert areas the sand has first to be bound together with grass, brushwood or nylon matting to stop it moving before being consolidated and irrigated, a process which can take a long time.

As the world population continues to grow, increasing use will be made of reclamation techniques to provide land for agriculture, industry and housing.

Lighthouses

Until the introduction of lighthouses the usual way to mark the entrance to harbours or the presence of rocks or sandbanks was by means of a beacon, made up of a pile of stones or wooden spars. These simple warning devices are still used in small harbours, and more elaborate structures such as masts and pillars are widely used.

The first lighthouse of which there is any record was the Pharos of Alexandria (in Egypt), a huge structure built by Ptolemy in the third century BC. It has been estimated that the tower was built on a base 100 feet (30.5 m) square and was 450 feet (137 m) high. The Pharos survived until about AD 1200, when it was destroyed by an earthquake; it gave its name to *pharology*, the science of lighthouse building.

The Romans built many notable lighthouses, such as that at Ostia (the chief port for Rome) and others in Spain and France and at Dover. Following the fall of the Roman empire, navigational aids, along with many other aspects of that civilization, fell into disuse. Once again it was left to individual ports to set up and maintain their own lights. One drawback was that a lighthouse is undiscriminating in its aid to ships, and a port with a lighthouse was an easy target for attacking ships. This restricted the use of lighthouses to peacetime, which hampered their development for many years.

From about the eleventh century onwards the increase in sea trading led to a revival of interest in lighthouse construction. Progress was slow, but in England and Europe from about 1600 onwards there was an increase in lighthouse building culminating in the great era of lighthouse construction in the eighteenth and nineteenth centuries.

Illuminants

The most important feature of a lighthouse is its light, and the efficiency of the lights has increased as technology has advanced. The first lights burned wood, which had the drawback of burning too quickly. Coal and candles were also used, but coal gave off so much smoke that soot collected on the lantern panes and blocked the light. The problem with candles was that it was difficult to produce a satisfactory level of illumination no matter how many candles were used.

Lighthouse illumination did not become really efficient until the early 1780s when the Swiss engineer Aime Argand invented the type of oil burner which bears his name. This lamp used a circular wick, surrounded by a glass chimney which created a central, upward draught of air to assist the burning. This lamp produced a steady smokeless flame of high intensity, and it remained the principal source of light for over 100 years.

The Argand burner was adapted for use in domestic gas lighting and gas lighting technology in turn contributed to the next major advance in oil burners for lighthouses. The Argand lamp used a wick from which the oil vaporized for burning, but in 1901 Arthur

Kitson produced a burner in which the oil was vaporized in a copper tube placed above the mantle (adapted from a gas mantle). The vapour then passed from the coiled tube to be burnt in the mantle, like gas. The oil was vaporized by the heat from the mantle, and a blowlamp was used to heat the coil before the lamp was lit. The Kitson design was improved by David Hood in 1921, and this type of burner is still widely used today where electric lighting is not practical. Many unattended lights burn acetylene gas.

Electric lighting was first tried in the South Foreland light on the Kent coast in 1858, using a carbon arc lamp, and this experiment was followed in 1862 by the installation of arc lamps at the Dungeness light, also in Kent. Arc lamps did not, however, prove satisfactory for lighthouses and little use was made of them.

The first use of electric filament lamps was also at the South Foreland light, in 1922, and many lighthouses are now operating with electric lights, either filament lamps or high pressure xenon lamps. The power is supplied from the local mains where possible, or else by diesel powered generators.

The third and most famous Eddystone light was designed by John Smeaton and completed in 1759, using a hydraulic cement invented by Smeaton, and with the stone blocks dovetailed together for strength. Erosion of the rock on which it was built necessitated its replacement by the present light which was built nearby, using an even more complex dovetailing arrangement than Smeaton's, which was dismantled and re-erected on Plymouth Hoe as a memorial to him.

Where a suitable rock foundation is unavailable, such as where the hazard to be marked is a sandbank or coral reef, the light may be built on steel piles or concrete filled caisson foundations. The light towers of these are often of open steel frame construction.

Lighthouses are usually equipped with some form of siren or horn which is sounded in foggy or misty weather, and may be controlled by an automatic electronic fog detector. In some manned lighthouses explosive fog signals are still used, typically going off at five minute intervals.

Light vessels

The first light vessel (lightship) to go into service was moored near the Nore buoy in the Thames Estuary in 1731, and was found to be of great benefit by ship owners, who willingly subscribed to her upkeep. The lighting consisted of two ship's lanterns mounted 12 feet (3.7 m) apart on a cross beam on the single mast.

The light vessel was put on station by Robert Hamblin, who thought that navigation suffered from the difficulty of distinguishing one lighthouse from another, and considered that light vessels should be moored at dangerous points around the coast, using different arrangements of lanterns to enable each station to be easily identified.

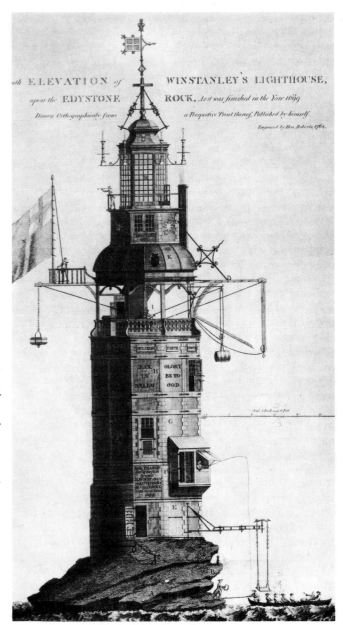

Right top: the original Eddystone light. Henry Winstanley, the designer, was killed when it was swept away in a storm in November 1703.

Left: the Dungeness lighthouse, built in 1960. It has 21 interlocking precast concrete rings, reinforced top to bottom by steel wires under tension. The honeycomb structure houses an electric foghorn.

Right: the mechanism for rotating the optics at the South Stack lighthouse at Anglesey, North Wales. The lens system itself is pictured overleaf.

The Nore vessel was placed on station against the wishes of Trinity House, the England lighthouse authority, who considered that lights which were only candles or oil lights would be ineffective as a guide to shipping. Despite these objections the King granted a patent for the light vessel, to run for 14 years from July 1730.

Following the immediate success of the Nore vessel, Trinity House became worried that light vessels would become so numerous as to upset the lighthouse system and eventually succeeded in persuading the King to revoke the patent in May 1732. The vessel had proved such a success, however, that it was impossible to remove it. Trinity House therefore obtained a patent in perpetuity and granted a lease for 61 years to Hamblin. After this initial breakthrough light vessels became accepted as valuable aids to navigation, but no other country tried them until 1800, and the first to be used in the USA did not go on station until 1820.

Optical systems

The development of efficient light sources led naturally to the development of reflector systems, because without some form of beam projection much of the intensity of the light is wasted. The three main groups of optical systems used in lighthouses are the catoptric (reflective), dioptric (refractive) and the catadioptric (reflective and refractive).

The first parabolic reflector was designed in 1752 by William Hutchinson, made up of small squares of mirrored glass set in plaster of Paris. The parabolic reflector is placed behind the light source and the rays of light are reflected parallel to the axis of the reflector and emerge as a beam of light. Reflectors made from a hand-beaten composition of copper and silver soon replaced the heavy glass reflectors, and by 1800 they were standard equipment in lighthouses.

The use of a reflector increased the power of the light signal by about 350 times, and as the problem of beam projection had thus been overcome the question of individual light characteristics for each lighthouse could be dealt with. This question arose because ship owners often complained that although lighthouses were useful, it was difficult to tell one from another as they all emitted the same signal.

This problem was solved by arranging reflectors in different positions on a frame and revolving it, producing group flashings (two or three flashes in quick succession, followed by a period of darkness, then the flashes again).

The most important development in lighthouse engineering was the Fresnel lens, invented by Augustin Fresnel in 1822. This is a dioptric lens, and has a central 'bullseye' lens surrounded by concentric rings of prismatic glass, each ring projecting a little way beyond the previous one. The overall effect of this arrangement is to refract (bend) into a horizontal beam most of the rays of light from a central lamp. Further reflecting elements may be placed above and below the refracting prisms to form a catadioptric arrangement.

Sometimes two lenses are placed one above the other, with a light at the centre of each, and this is called a bi-form optic. Many refinements have been made to Fresnel's original design, but the basic principle is essentially the same today.

Improved methods of producing and finishing glass, and the development of plastics, have made it possible

to reduce the size and weight of optical systems. This, together with the improvement in light sources, has enabled more efficient and compact apparatus to be produced.

Construction

Most lighthouses are built of stone or precast concrete, and some are built many miles off shore. The Eddystone light, for example, stands on a rock in the English Channel some 13 miles (21 km) from Plymouth. The present structure is the fourth to be built on the site, and was completed in 1881. The original was a wooden tower, built in 1698 by Henry Winstanley who was at the light in 1703 when it was washed away during a severe storm. The second one, also of wood, was built by John Rudyerd in 1708 and survived until 1755 when it burned down.

Above: the lamp and part of the lens system at the South Stack lighthouse. The circular parts are Fresnel lenses.

Left: one of the four lightships which mark the Goodwin Sands, a dangerous stretch of shifting sands near the straights of Dover. The sands are often exposed at low tide, and ships have run aground despite lights and markers.

Right: a cross-section of an offshore lighthouse built of interlocking stone blocks, with a more detailed section of the service room and the lantern gallery.

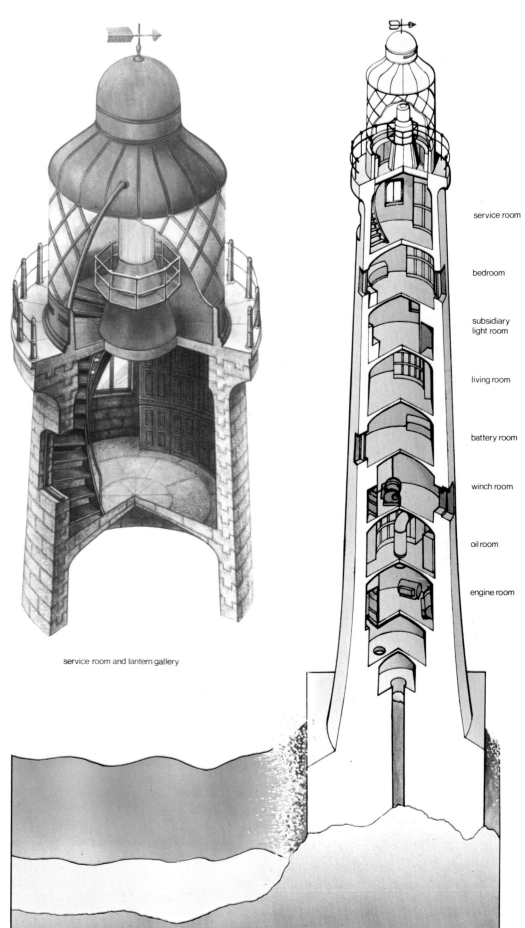

service room and lantern gallery

service room

bedroom

subsidiary light room

living room

battery room

winch room

oil room

engine room

Mining und quarrying

Not until the application of steam power to ventilation, drainage and hoisting were any really significant developments in mining techniques made. The industrial revolution of the 18th and 19th centuries brought about more progress in mining than had been achieved in the previous 5000 years of mining history. This transformation did not take place without a heavy price being paid. As mines and mining became more complex so did the toll in human suffering, and the world's mining industry entered the present century with an estimated annual death toll of the order of 50,000.

In the first half of the 20th century mechanization paved the way for booming production but replacement of men by machines was not always an unqualified blessing, and often this swing away from pick and shovel mining meant that the accident rate was increased. However, stricter safety standards and improved machines and techniques introduced over the past several decades have brought about a significant improvement in overall safety. Despite the fact that the total yearly tonnage of mineral and ore now mined is several times greater than that produced at the beginning of the century, the world accident rate has been reduced by some 90%. Nevertheless mining is still inherently a dangerous occupation, and until such time as machines take over entirely from miners, a continuing price in human suffering will

unfortunately be inevitable.

The approximate current annual world production of various minerals, ores and rocks given in millions of tons includes: metallic ores (2000), non-metallic ores (1400), road and building stones, sand and gravel (3500), and coal (3000).

Surface operations account for almost 70% of all mining production. When it is borne in mind that in some surface operations, in order to expose the deposit, the mass of overlying rock to be moved may be 10 to 20 times greater than the tonnage of the deposit itself, it is easy to appreciate that in world mining something like 50,000 million tons of material (including waste) are moved annually. The scale of individual mining operations varies tremendously. Mines range in size from operations employing two or three workers and producing a few thousand tons of ore a year to giant South African gold mines, each employing 15,000 men and winning many millions of tons of gold ore each year from depths up to 12,000 ft (3.7 km) below the surface. At such great depths the air passed through the workings must be cooled in giant refrigeration plants because of the high rock temperature.

The world's largest underground mine is the huge Kiruna iron ore mine in Swedish Lapland with an annual output of 15,000,000 tons of high grade magnetite. Top

Left: a slate mine in North Wales. The typical terraced look of an opencast mine makes it easier for trucks to get into it.

Right: Mount Morgan mine, Queensland, Australia. In 1973, 8488 metric tons of copper, 53,432 ounces of gold and 42,882 of silver were produced from 1.3 million tons of ore.

Above right: the small picture gives some idea of the size of an excavator used for stripping overburden. These are among the largest machines on earth.

of the production league in surface mines is the Bingham Canyon copper mine in Utah, USA. Here almost 250,000 tons of waste rock are stripped each day and 90,000 tons of copper ore having a copper content of 0.7 % are mined daily. Although this is called a surface mine the floor of the open pit is more than 2000 ft (0.6 km) below the surface.

Mining methods

Quite apart from the desirability of reducing the number of men at risk in mining operations, the high wages cost of labour in most mining areas has dictated the replacement of men by machines in the performance of mining operations. Equipment has been evolved which will win, load and transport minerals and ores with the minimum of manual attention.

It is relatively easy to mechanize operations at the face where the material is comparatively soft, like coal or phosphate. Machines have been devised which cut out the roadways underground and other machines have been developed which automatically cut and load mineral on longwall faces. Hard rocks, however, such as metalliferous ores, are not amenable to cutting and it is difficult to mechanize the actual process of breaking down the ore from the parent mass underground because of its hardness and the irregular shape of many deposits. Thus most underground metalliferous mining methods involve dividing the ore body into working sections (called *stopes*) and subsequently employing explosives to blast the ore away from each working face exposed by the underground development tunnels.

In the blasting processes, holes about 1 to 2 inches (2.5 to 5 cm) in diameter are drilled 6 to 10 ft (2 to 3 m) into the face of the ore Explosive charges are fed into these holes and then detonated, so bursting away the ore ready for loading into transport media for despatch to the surface. Although the amount of explosive varies widely with conditions, an average 10 ft (3 m) hole charged with 4 lb (2 kg) of explosive would yield about 3 tons of ore. Sometimes many explosive charges are detonated simultaneously or in rapid sequence and the use of a ton or more of explosive in one blast is not uncommon. Blasting is usually done between shifts so that men are not endangered and sufficient time is given for noxious fumes to be dispelled.

Although alluvial mining of tin has been practised for a century or more, the principle involved has been considerably developed in recent years. The technique involves the dredging from the sea bed of a multiplicity of valuable products ranging from diamonds to the more prosaic—but very important—sand and gravel deposits. These unconsolidated deposits are won by bucket wheel, grab or suction pump dredgers, and although most commercial marine mining operations are confined to a maximum water depth of about 200 ft (60 m) or so, prototype air-lift pumps have been used to recover manganese nodules from the floor of the Pacific Ocean at a depth of 3000 ft (900 m). Some authorities believe that economic mining of the sea bed is possible down to 12,000 ft (3.6 km) depth and that the tonnage of valuable metals thus available exceeds total land reserves.

Above: preparations for blasting at a mine in Peru. The hole has been partly filled with ammonium nitrate and fuel oil, and a packing layer of sand is being poured in.

Right: ore crushers at an Australian iron mine.

Another specific surface mining technique is borehole mining. This involves the sinking of boreholes from the surface. In the Frasch process for the mining of sulphur, boiling hot water is passed down pipes placed in a borehole. This water melts the sulphur, which is then pumped to the surface as a liquid. Another form of borehole mining involves passing a solvent down the borehole. This liquid dissolves the valuable deposit and the solution is then pumped back to the surface. This system is most commonly applied in salt mining (using water as the solvent) but solution mining has, however, been used to selectively dissolve other minerals such as potash. On reaching the surface, the impregnated solution is treated and the valuable mineral regained.

In the mining of some relatively soft surface deposits such as China clay, high pressure water jets (up to about 100 psi or 7 bar) are used to dislodge the valuable clay, which is then sluiced down to the treatment plant. This is known as hydraulic mining. This principle has also been employed to a limited extent in underground coal mines in Europe, in particular in the USSR.

There are various techniques for winning ores underground, depending on the extent, shape and geology of the deposit. Where large masses occur caving may be used; here tunnels are made in the mineral in such a way that, on blasting, it literally caves in from above, falling to the collecting level below. Alternatively, in stoping, the ore is systematically worked from various levels—a method suited to both mass and vein deposits. This also includes 'room and pillar' method, where various chambers are excavated by leaving pillars between to support the roof.

In metalliferous mining, a system which finds some application is the leaching technique. This technique incorporates principles of hydrometallurgy and essentially involves passing a weak acid over the broken ore underground. This acid selectively dissolves the metal from the broken rock and is then pumped back to the surface. In many respects in situ leaching is similar to the borehole mining previously described, the essential difference being that only a very small part of the deposit is dissolved—only the metal content, and this may be just a fraction of 1 % of the mass of the ore deposit. The chief difficulty with in situ leaching is that almost invariably the ore mass must be broken up by normal mining operations; otherwise, the weak acid cannot attack sufficient ore.

Coal mining

Many hundreds of years ago, Europe and the British Isles were extensively covered with forests, but in modern times wood as a source of fuel has become comparatively scarce and therefore expensive. The discovery and exploitation of coal has had an important economic motivation. About 1200 AD a monk near Liège (in Belgium) made reference in a chronicle to a black earth similar to charcoal used as a fuel. From the 13th to the 16 centuries the area around Liège made the most extensive use of coal in the world. The town of Liège dug its mine shafts in hills overlooking the town and drained them in such a way as to obtain its main water supply from them, thus deriving more than one advantage from the mining operation.

Coal outcrops had been exploited in England since the Middle Ages. In the time of Elizabeth I, London began

to import coal from mines in the northern part of England; from then until the late Victorian period Britain had no near rival in its production of coal. London's factories and homes produced so much coal smoke that it quickly became the dirtiest city in the world. The poisonous, acrid fumes combined with climatic conditions finally became a health menace which led to the necessity for the Clean Air Act of 1956.

The switch from wood to coal that occurred in the 17th century established the foundation of the Industrial Revolution which followed. A coal fire is so dirty compared to a wood fire that a whole new technology had to be developed in many industries in order to deal with it. For example, in the breweries when coal fires were first used to dry the ingredients the resulting beer was undrinkable. The use of coal also encouraged the beginnings of modern mass production: in the glass industry coal fires made possible greater production of plate glass than

Left: a 'cactus grab' sinks an air shaft in a mine in Australia producing lead, silver and zinc.

Below: strip mining is much cheaper.

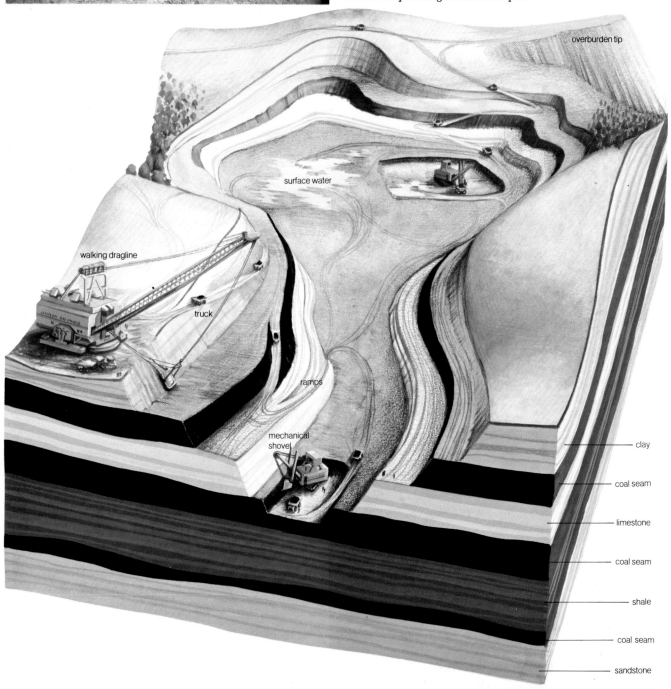

overburden tip

surface water

walking dragline

truck

ramps

mechanical shovel

clay

coal seam

limestone

coal seam

shale

coal seam

sandstone

ever before, but the creation of beautiful objects one at a time by glass blowers became much more difficult in the smoky, poisonous atmosphere.

The first underground workings consisted of little more than vertical shafts some 6 ft (1.8 m) in diameter sunk to a depth of 30 ft (9.1 m) or so. The coal around the bottom of the shaft was then hewn out to a radius of several feet and pulled or carried manually in baskets up the shaft. Such diggings are called bell pits, and post-war opencast mining operations in Britain have uncovered several of them.

As mines went deeper and became more complicated, technological progress was made only at the cost of human life. Each advance in the daring of the miners led to disasters against which safeguards were then developed. One early solution to the problem of ventilation was to sink a parallel shaft deeper than the one being worked. Birds were taken into mines because they use oxygen faster than humans, and if the air supply became inadequate the birds would die soon enough to give an early warning. Another problem was the flammable methane gas which is given off by coal seams when they are exposed by digging. In the early days of mining, a man would wrap himself in wet rags and crawl along the floor of a shaft, holding a burning torch on a pole above and ahead of him to ignite the gas, whereupon the mine would be considered safe. Today safety methods are considerably advanced compared to these, but even so disasters taking many lives have occurred in the 20th century.

Surface mining is carried out by stripping away the strata overlying the coal seams and then removing the exposed coal. Until comparatively recently it was only feasible to remove a maximum of about 100 ft (30 m) of surface strata (overburden) but the post-war years have seen the development of huge excavating machines capable of stripping several hundred feet of overburden. The loading buckets on such machines are capable of scooping up several hundred tons with each operation. Such machines can achieve hourly loading rates measured in thousands of tons. Not only has the size of the machine increased but there has been a general increase in the scale of opencast operations, and some surface mines are now producing up to 50,000 tons of coal a day.

For use under certain favourable conditions, a special surface mining technique known as auger mining has been developed in the USA. The method involves boring a series of parallel holes into coal seams which have out-cropped (been exposed by removing overburden). These augers are simply large rotary drills 2 to 5 ft (0.6 to 1.5 m) in diameter which bore into the coal seam for some 300 ft (91 m). The coal which is cut away by the auger travels back along the scroll of the drill rod and is collected at the mouth of the hole. Augering machines are usually used where the thickness of overburden becomes too great for further removal, but they can sometimes be used underground as well. In circumstances where it can be used the auger machinery provides the cheapest coal mining methods of all.

To produce one ton of coal it may be necessary to strip as much as 30 tons of overburden, which provides some concept of the size of a modern open pit. In such pits the removed material may be loaded onto trucks capable of hauling 100 to 200 tons, or alternatively onto conveyers which can move 10,000 tons an hour. The productivity of miners in open pits is very high compared to under-

ground workers and outputs of 50 tons a manshaft are recorded.

Total mining costs are low by underground standards and may be only one quarter of those in deep mining operations. For these reasons opencast mining is favoured where possible. In Britain less than 10 % of national output is opencasted but in the USA nearly half the output is mined this way.

Opencast or strip mining has become an emotional issue among conservationists, particularly in the USA. In some places whole tops of mountains are being cut off. This affects not only the natural beauty but also the drainage of the land, wildlife habitation, and so on. One solution would be to put some of the overburden back and replant it when the surface mine is exhausted, possibly using rubbish collections brought from the cities as additional landfill.

Most of the world's coal is still won from underground mines, some of which are 3000 to 4000 ft (900 to 1200 m) deep. In such depths, access to the seams is by vertical shafts equipped with hoisting machinery, but in shallower depths down to 1000 ft (300 m) the workings may be connected to the surface by inclined tunnels. Conveyers are usually installed in these surface slopes.

There are two principal methods of underground working: *room and pillar* and *longwall working*. With the former system, once access to the seam has been gained, tunnels (rooms) are driven into the seam in two directions at right angles so dividing the seam into a number of rectangular blocks of coal (pillars) which may or may not be subsequently extracted. Depending on certain practical

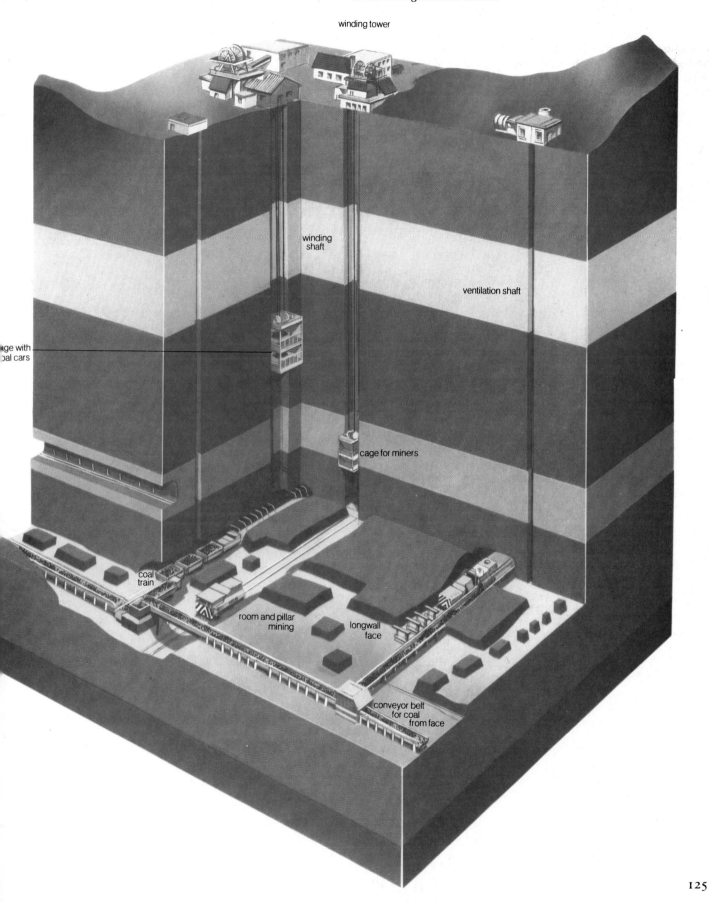

Opposite page: mining history is a story of drudgery and danger. Children and ponies were used to push cars of coal up inclined shafts (1842). In the other engraving (1869) the 'penitent' or 'fire-man' ignites the methane gas given off by the coal seam.

Below: an underground coal mine. Proper ventilation did away with the need for birds in cages, which died if the air went bad, giving warning to the miners. The room and pillar method is illustrated; if the pillars are removed, the method is called longwall extraction.

winding tower

winding shaft

ventilation shaft

cage with coal cars

cage for miners

coal train

room and pillar mining

longwall face

conveyor belt for coal from face

considerations such as the degree of roof support needed from the pillars and the type of machinery being used, the rooms are 9 ft to 24 ft (3 to 7 m) in width and the intervening pillars from 30 ft (9 m) square to 150 ft × 300 ft (45 × 90 m). Machines have been developed for driving these tunnels which eliminate the need for manual breaking or shovelling of coal. Such machines can cut tunnels in the seam at speeds of up to several inches a minute to produce coal at a rate of up to 10 tons a minute. This coal is mechanically taken to the rear of the machine and loaded onto conveyers or wagons for transport to the surface. Room and pillar working is favoured when mining beneath surface buildings or under lakes and seas. Under such circumstances the pillars are left in position to minimize movement of the ground at the surface. When the pillars are left, the term *partial extraction* is applied to the system.

Longwall working is a total extraction system: all the coal within a specified area is extracted in one operation. Two parallel tunnels are driven into the seam some 150 to 600 ft (45 to 180 m) apart. These tunnels (gate roads) are then joined by a road at right angles, this third road forming the longwall face. Successive strips are then taken off the side of the face road and the coal is deposited on a face conveyer which delivers it to the gate road conveyer and the shaft. As the longwall face moves forward, the roof behind the face is allowed to collapse, the gate roads being correspondingly advanced and supported. Such faces can advance several yards a shift and produce a daily output of 7000 tons. Many longwall cutter operations have been automated, as have the accompanying roof support systems. It is now possible for all operations to be performed by one operator situated well back from the production area. The most advanced systems include a nucleonic probe, an automatic steering device which sends a radioactive low-frequency pulse into the seam above and below the machinery, guiding the cutting operation along the seam and away from the denser surrounding rock.

These highly automated systems can cost hundreds of thousands of dollars. The daily output, however, may be several thousand tons, and the effective costs per ton will be less than for manually operated production units.

Productivity in underground mines varies widely, being so dependent on mining conditions; overall output ranges from 1 to 10 tons per manshift.

The underground gasification of poor quality coal seams is extensively used in certain coal mining areas, although limited trial installations in Britain and Western Europe have not proved too successful for various technical reasons. Broadly the system involves sinking shafts to the coal seam and then igniting the coal underground and piping the evolved gases to the surface. Such gases have a low calorific value—only about one fifth of that of natural gas—and thus it is not economical to pipe the gas a long distance on the surface to the point of utilization.

Quarrying

Quarrying is the removal of rock from the Earth's crust for use in the various construction industries.

The three types of rock are *igneous*, *sedimentary* and

An opencast ('strip') mine near Cologne. 100,000 square metres of lignite are produced here each day.

A 'robot' taking coal in a mine in Belgium. As fossil fuels go up in price, more automation can be used.

A mechanical scraper removing overburden in the Brunswick strip mine, Helmstedt. A similar machine is pictured on page 70, but in this photo the size of the man gives the relative size of the machine. On the opposite page there is a view from a distance of the mechanical scraper. These devices, along with the dragline excavators (see pp. 66 and 71), are among the largest machines on earth.

A pit head at Ebbw Vale, Wales. The wheels and cables haul the 'cage' up and down.

127

metamorphic. Igneous rock is formed under conditions of great heat and pressure; granite is a good example. Sedimentary rock is a mechanical agglomeration of particles; examples are sandstone and limestone. Metamorphic rocks, as the name implies, are combinations of materials, largely igneous but including others and requiring special circumstances for their formation.

Sedimentary rocks are called *fragmental*; in between igneous and sedimentary rocks are groups of rocks called *pyroclastic* (volcanic) which are also called fragmental volcanic.

For the purpose of quarrying, rocks are divided into two types: dimension stone and crushed stone. Dimension stone is cut from the Earth in blocks; granite is the best example. It is suitable for both interior and exterior facing in building construction and for monuments and memorials. Sandstone is also made into building blocks, and into abrasive grinding wheels. Slate is manufactured into waterproof roofing as well as chalkboards. Crushed stone is universally applied for drainage and ballast in road building and under railway tracks, and as aggregate in concrete. There are also many special uses for crushed stone; some types can be used in suspension in abrasives or with bonding agents to make grinding wheels; crushed limestone has many chemical applications and is used as fluxing stone in blast furnaces; in paper manufacture, chalk is a whitening agent and makes cigarette paper burn evenly.

In ancient times, quarrying was almost entirely of dimension stone; ancient peoples such as the Egyptians performed amazing feats of quarrying and transporting great blocks of stone without modern tools and without certain principles of physics, as embodied in the block-and-tackle. In modern times the tonnage of crushed stone far outweighs that of dimension stone because of the widespread use of concrete in building construction, the tremendous increase in road building since the Industrial Revolution (as well as the technological advances in road building required by heavy traffic), and the industrial and chemical applications of crushed stone.

In quarrying, the first step once the rock deposit has been located is to clear away the soil, gravel and so forth from the rock face, a process similar to the first stage of opencast or strip mining. Gravel itself is an important aggregate, comprising more-or-less rounded particles larger than grains of sand. The rounding is accomplished by abrasion by water, and gravel beds are found near beaches and in pre-historic river beds. They often contain deposits of mineral ores. Not much quarrying technology is required in a gravel bed.

Crushed stone

Once the rock has been exposed, the next step is to drill holes in rows. *Churn* or *well* drills, which are cable or rope operated tools called cable-tool drills in the USA, are used. The holes are about 6 inches (15 cm) in diameter and 50 ft (15 m) or more deep. The holes are drilled into the rock, usually 20 or 30 ft (6 to 9 m) back from the rock face; the distance is called the burden. The spacing between the holes is usually less than the burden. There may be twenty or more holes and more than one

row, depending on the size of the operation. Explosives are packed into the holes, and the top part of the hole is packed with sand or dirt. The size of the explosive charge depends on the type of rock being quarried. As much as 20,000 tons of broken rock may be blasted loose in one operation.

The rock is then crushed in machines. These may be cone, jaw, hammer or roller crushers, with ratios of feed size to crushed size ranging from 3:1 to 10:1; the crushing may be done in several steps with screening of the rock between each stage. Crushing to very fine or powdery consistencies is called pulverizing or disintegration.

Dimension stone

Deposits worth quarrying for dimension stone are chosen on the basis of colour and grain of the rock, and particle size and uniformity. Obviously, since the object is to win large blocks of stone, explosives are not used. In addition, blocks must be removed carefully to preserve their quality and to avoid cracking them. Advantage is taken wherever possible of the natural joins in the rock formation.

The first step is to make vertical cuts in the rock. A channelling machine is used for softer rocks; this comprises several steel bars with chisel edges clamped together. It cuts a channel several feet deep and about 2 inches wide (5 cm). For harder rock such as granite, the most common method is the wire saw. This is a continuous wire rope about 3/16 inch (about 5 mm) in diameter, which is fed with abrasives suspended in water as it runs.

If there is no convenient natural join at the base of the

Quarrying, along with mining, is one of Man's earliest industries. In fact, a quarry is really a kind of opencast mine.

Above left: sawing Purbeck stone at St. Aldhems Quarry. A six-bladed cross-cut saw is used; water mixed with abrasive cuts the stone and cools the blades.

Above centre: the Carrara marble quarry in Italy, said to have been Michelangelo's quarry. Large pieces of marble are separated first and then cut into smaller pieces on the site.

Above right: an open-pit quarry in Cumberland. The quarrying of crushed stone nowadays is many times the volume of dimension stone quarrying, because of its use in roadbuilding and as aggregate in concrete, and because concrete and steel are used in building where stone would have been used in former times.

block, it must be carefully separated from the bed. Horizontal holes are drilled, and wedges used to break the block loose. The blocks broken loose are often very large, as much as 100 ft long (90 m). They are divided into smaller blocks on the spot, usually by the plug-and-feather method. Holes are drilled, and feathers (tapered lengths of iron, round on one side) are inserted two to a hole, and the plug, a steel wedge, is driven down between them. The blocks are then taken to mills where they are further sawn, shaped, or turned on lathes into columns, and polished.

Pyramids and stone circles

It is fascinating to speculate about the construction details, as well as the original purpose, of ancient building projects which survive today. The dedication and the man-hours which went into these monuments was truly extraordinary.

The Pyramids

Egyptian pyramids were the dominant features in complexes of buildings intended to serve the needs of important persons after death. They are of two kinds: step pyramids, in which all four sides are built in steps from the base to the summit, and true pyramids, in which the sides slope continuously (normally at an angle of about 52°) from bottom to top. Step pyramids are the older: the first was built for King Zoser of the Third Dynasty (c 2660 BC) under the direction of his famous architect Imhotep. True pyramids superseded step pyramids at the beginning of the Fourth Dynasty (c 2600 BC) and kings continued to build them for about 1000 years. About thirty of these pyramids have survived, though many of them are in a ruinous condition. No two pyramids, whether step or true, are exactly alike either in size or in internal design.

Four important considerations governed the choice of a site for building a pyramid. It had to be located near the royal residence, on the west side of the Nile, out of reach of the annual flooding and at a place where the rock was firm and fairly level. Surface irregularities were probably removed by a laborious process which consisted of erecting a low wall around the perimeter of the site, flooding the whole of the enclosure thus formed to a depth

of a few inches and cutting a network of trenches with their floors at an even depth beneath the surface of the water. The wall would then be dismantled, releasing most of the water immediately, and the remainder would soon evaporate. Lastly, the uncut rock within the network would be reduced to the level of the floors of the trenches. In the Great Pyramid, apparently exceptionally, a knoll of rock in the centre of the site was left standing and eventually embodied in the building. Nevertheless, the rest of the bed of the pyramid deviates from a perfectly level plane by little more than half an inch.

Immense care was taken to orientate the pyramid so that each side would directly face one of the four cardinal points. The only method known to the ancient Egyptians by which the accuracy displayed in the major pyramids could have been achieved was by setting a rod vertically in the rock and sighting from it the rising and setting positions of a star on the eastern and the western horizons. A line drawn to the rod from the midpoint between those two positions would run directly north–south. The east–west axis could easily have been determined by using a set square, an instrument which the ancient Egyptians are known to have possessed. Since any rise in the ground would have prevented an observer from seeing the true horizon, a circular or semi-circular wall, level at the top, was probably built, the rod being placed at its centre and the sighting made as the star appeared above the artificial horizon provided by the wall and as it disappeared beneath it.

Composition

Pyramids which are sufficiently damaged to enable their composition to be examined show that the inner corner was built in upright layers around a central nucleus. Each layer in a true pyramid normally consisted of horizontal courses of local stone faced with an outer skin of fine quality limestone obtained from the quarries at Tura in the Maqattam Hills, on the east side of the Nile opposite Giza and Saqqara. The outer sides of the nucleus and of the surrounding layers inclined inwards at an angle of 75° and thereby greatly reduced the risk of any tendency for outward pressure from the weight of masonry. Nothing was done to smooth the inner and the outer surfaces of the blocks of Tura limestone used in the core. Chambers and corridors, if they were located in the superstructure, were built as the monument was being erected and large objects such as sarcophagi (stone coffins), if they were of larger dimensions than the corridors, were placed in position before the roofs were built

Left: Zoser's pyramid was begun as a mastaba, a rectangle with a flat top. This was extended sideways and given progressively smaller top pieces.

Above: the main pyramids were built at a time when the only metal available for tools was copper. The blocks were quarried by cutting slots with picks, then driving in wedges. Pick marks can still be seen at Aswan.

Right: this ruined corner of the Bent Pyramid at Dahshur shows the blocks on the inside and the smooth outer cladding. It changes slope halfway up.

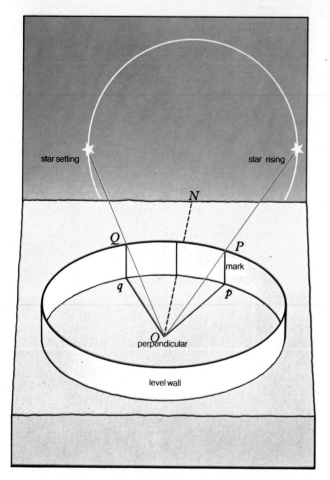

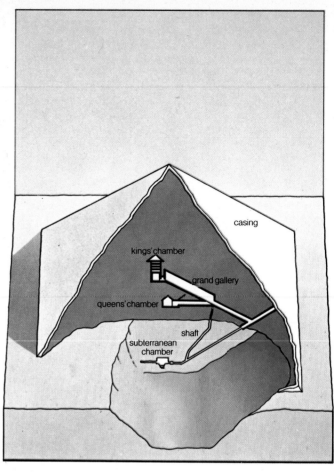

A mark is made on the wall (P) where the observer (O) sees a star rising, and at Q where it sets. A line drawn from the centre point (pq) to O will run due North and South.
The other picture shows a cross section of the Great Pyramid. No two pyrimids, whether step-sided or smooth-sided, are alike either in size or internal design.

over the chambers. Subterranean corridors and chambers were tunnelled in the rock before any masonry was laid.

At the same time as the construction of the core, whose upright layers diminished in height from the centre outwards and thus resembled a step pyramid, the whole pyramid was encased with an outer skin of Tura limestone. Each casing block was cut before laying to the pyramid angle of about 52°. The Great Pyramid slopes at an angle of 51° 51′, an angle which would have resulted from a plan to make the height equal to the radius of a circle whose circumference was equal to the perimeter of the pyramid at the base. Egyptian mathematical texts provide no evidence for supposing that the true nature of π was understood—it is impossible to represent by a fraction. The last stone to be put in position, the capstone, was a miniature replica of the pyramid itself, generally made of granite and sometimes overlaid with gold. It was held firmly in place by a tenon projecting from its base which fitted into a socket cut in the uppermost course of the masonry.

It has been estimated that the Great Pyramid consists of about 2,300,000 limestone blocks of an average weight of $2\frac{1}{2}$ tons and reaching a maximum of 15 tons. In addition, much granite from Aswan, 500 miles (800 km) south of Gizeh, has been used in its interior; the nine slabs forming the ceiling of the King's Chamber alone weigh nearly 45 tons each. Not only were a large number of these blocks loaded on ships and brought to a quay near the site by river, but they and many of the blocks quarried locally had also to be moved on land to the immediate vicinity of the pyramid and ultimately raised to the position which they were to occupy in the superstructure.

Transport was by means of wooden sledges, pulled by man-power over temporary roadways paved with baulks

of timber. There is no positive evidence of the use of rollers, but it is possible that they were employed both under sledges and under the blocks when they were not mounted on the sledges. Only one method of lifting stones in building was practised by the ancient Egyptians before the introduction of the pulley in Roman times, and that was by hauling the block (on its sledge) up a brick and earth ramp, the upper end of which leant against the part of the building already finished. In the case of a pyramid such a ramp would have been necessary on one side only. As the pyramid grew in height, the ramp would be raised and extended in length, so that the same slope could be maintained. The remainder of the pyramid would have brick embankments of sufficient width to allow the builders to stand on them and lay the outermost stones in position. When the capstone had been laid, the embankments and the ramp would be dismantled and as the work proceeded the outer faces of the Tura limestone casing, which until then had remained rough, were smoothed from top to bottom.

One of the many striking features in the building of the pyramids is the perfection achieved in laying the stones of the outer casing. Sir Flinders Petrie estimated that the mean thickness of the joints in the Great Pyramid was 0.02 inch (0.5 mm). Mortar was used mainly as a lubricant to enable the masons to manoeuvre the stones while laying them.

Stonehenge

The great prehistoric monument of Stonehenge on Salisbury Plain has for centuries been regarded as one of the wonders of Britain. In the 12th century it was thought to have been created magically by Merlin, King Arthur's magician; later antiquaries saw it as a temple of the Druids, the Celtic priesthood at the time of the Roman invasion of Britain. Modern archaeology shows that Stonehenge was already over 2000 years old at the time of the Celts, and that there is no evidence that the Druids were connected with it at all, but the debate as to its original purpose is still being settled.

Archaeological research shows that Stonehenge was built in three main stages. About 2800 BC the late New Stone Age farmers built Stonehenge I, consisting of a circular bank and ditch 320 ft (97 m) in diameter with a concentric circle of 56 holes 10 ft (3 m) inside. These Aubrey holes (named after the 17th century antiquary who discovered them) are 2 to 4 ft deep (0.6 to 1.2 m) and some contain contemporary cremations which allow dating by the radiocarbon technique. There was an entrance in the north-east part of the bank, and outside it was the Heel Stone, an unshaped standing stone which is still in place.

The arrival from Europe of the 'Beaker People', an early Bronze Age culture, led to the rebuilding of Stonehenge some centuries later. This Stonehenge II was a concentric double circle of 38 stones apiece, with an entrance facing the Heel Stone. This direction continued to be the 'axis' of Stonehenge throughout its subsequent reconstructions, and its astronomical importance is discussed later. The stones used in the building of Stonehenge II were brought from the Prescelly Mountains

The Rollright stones in Oxfordshire. A line from the centre to the King's stone marks rising of Capella.

in Pembrokeshire, 135 miles (216 km) as the crow flies. The route actually taken was probably along the Bristol Channel, up the Bristol Avon and then overland to the Wylye and the Salisbury Avon, a total distance of 240 miles (383 km). From Amesbury they were dragged overland for 2 miles up the 'Avenue' (a wide passage marked by two earthen banks) which led from the river to the entrance of Stonehenge. It is not known why these *bluestones* (so named because of their dark blue-grey colour) were brought from such a long distance; and, as it transpired, Stonehenge II was never completed. When three-quarters of the stones had been set up, they were removed to make way for the next stage.

It is the remains of Stonehenge III which impress the visitor today. The stones, up to 30 ft (9 m) long and up to 50 tons in weight were arranged in two concentric structures, the outer consisting of 30 uprights supporting a continuous circle of lintels about 14 ft (4.3 m) above the ground. The inner structure comprised a horseshoe arrangement of five *trilithons* (arches of three stones each), arranged with two on either side of the axis of the monument and the fifth and largest on the axis at the opposite end from the Heel Stone.

All the stones used in Stonehenge III, called *sarsens* from the word Saracen, meaning 'strange', were transported from the Marlborough Downs, 20 miles (32 km) north, where they occur as individual sandstone boulders on the chalkland. It has been calculated that it would have taken 1500 men ten years to move all 81 sarsens. From the holes in which the sarsens stand it is possible to deduce a certain amount about how they were erected. The outward face of each hole is not vertical, but slopes at about 45°, and it is thought that a sarsen was moved over its hole from this side until it overbalanced and the end fell into the hole. The stone would then be resting on the sloping face, and would already be half way to a vertical position. The final straightening may have been by levering it from successive layers of wooden packing inserted underneath to take the weight until it could eventually be pulled upright. Traces of vertical wooden stakes have been found against the inside faces of the holes, and these were probably positioned so that the foot of the sarsen would not scrape the soft chalk from the face into the bottom of the hole as it was being erected. Any accumulation of this kind would have affected the heights of the top of the uprights unequally, and hence the lintels would not have been level. The lintels were probably raised by levering them up on wooden platforms, as there is no sign of soil disturbance which would have occurred if earth ramps had been used.

This structure was built in about 2100 BC, according to the radiocarbon dating, and in the following centuries some of the bluestones used in Stonehenge II were re-erected within the central area, and in a circle within the sarsen circle. A massive sandstone slab, popularly known as the Altar Stone, which now lies flat at the centre of Stonehenge, was also part of Stonehenge III. It had been quarried at Milford Haven, in South Wales, and was presumably transported in the same way as the bluestones.

A brief archaeological description of Stonehenge gives

Opposite page: Stonehenge from the air. In 4000 years many of the stones have fallen, been moved or reset, but excavation shows the original holes in the chalk. Around the edge are the 56 Aubrey holes, now filled with concrete but originally with pressed chalk. At the bottom of the photo the 'slaughter stone' can be seen; named by Romantics it is actually one of a pair of portals.

Left: the mortise and tenon structure of the trilithons is clearly seen in the remains of the largest of all. The upright (left) is nearly 30ft (9m) tall and weighs 50 tons.

Above: modern druids, from a sect begun in the 18th century, hold a ceremony annually at Stonehenge. The Heel Stone leans slightly, but as the ground level is lower, the line still works.

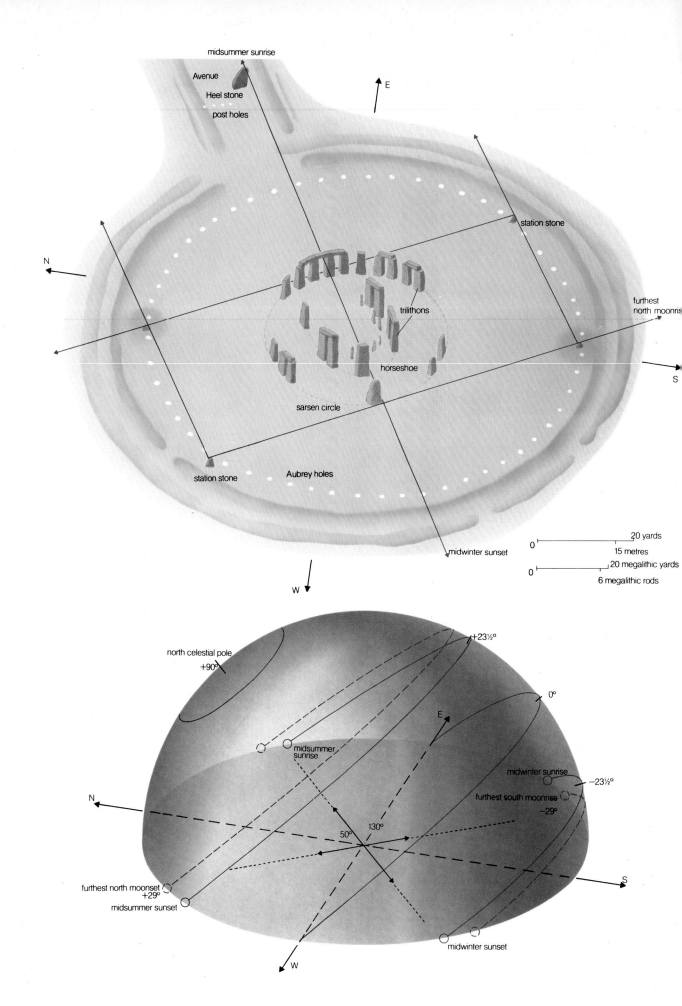

midsummer sunrise

Avenue

Heel stone

post holes

E

N

station stone

furthest
north moonris

trilithons

S

horseshoe

sarsen circle

station stone

Aubrey holes

midwinter sunset

W

20 yards
0
15 metres
20 megalithic yards
0
6 megalithic rods

north celestial pole
+90°

+23½°

0°

E

midsummer
sunrise

midwinter sunrise

−23½°

furthest south moonrise

−29°

N

50°

130°

S

furthest north moonset
+29°

midsummer sunset

W

midwinter sunset

Above: carved daggers on a Sarsen stone were seen as evidence that the builders of Stonehenge were influenced by the Mycenaeans, but radiocarbon dating disproves this. Opposite page: significant alignments at Stonehenge. (Compare sky sphere below.) It is impossible to know whether Bronze Age Man could predict eclipses, but he was certainly aware of the Moon's motion.

no explanation of its original purpose. The effort involved in its construction was immense, not only in transporting and erecting the stones, but also in shaping and smoothing the sarsens, a task which, it is calculated, would have taken 100 men 15 years of pounding with stone hammers. It is also unique in Northern Europe because all the touching stones are jointed together. Tenons on the tops of the uprights project carefully into matched hollows (mortises) on the undersides of the lintels; and the adjacent lintels in the outer circle are connected by tongue and groove joints. This type of carpentry technique applied to stone is unknown elsewhere north of the Alps at such an early period. At one time this gave rise to the speculation that the architect had experience of the Mycenaean architecture of Greece, but recently all radio-carbon dates have been revised and the new date for Stonehenge makes it considerably older than the Greek civilization: it seems certain that there was little foreign influence in its design.

Midsummer sunrise

It is difficult to understand why the Bronze Age people of Wessex thought the effort of constructing Stonehenge worthwhile, unless it was ordered by a powerful ruler or incited by religious fervour. For centuries it has been assumed that Stonehenge must indeed have been a temple, and because the axis of Stonehenge points over the top of the Heel Stone to the point on the horizon where the Sun rises on Midsummer's Day, it was natural to think that it was a temple for the worship of a Sun-god. To this day, visitors still flock to Stonehenge to watch the Sun rise on 21 June.

Recently it has been suggested that this alignment is not religious, but was constructed because Stonehenge was an observatory to study the movements of the Sun. The tilt of the Earth's axis of rotation leads to the Sun's apparent path in the sky changing during the year. It is well known that the height of the Sun at noon is greatest at midsummer and least at midwinter; and the points of sunrise and sunset move around the horizon corre-spondingly, particularly noticeable in high latitude regions such as Britain. At the midsummer *solstice* (21 June) the Sun rises and sets at its most northerly position, at Stonehenge about 50° around the horizon from true north, while at the midwinter solstice (22 December) it rises and sets 130° from north. The farmers of the New Stone Age and the Bronze Age would have needed a calendar to know when to sow their crops, and a natural way to fix the calendar dates would have been by reference to one of the solstices, when the Sun reaches its most northerly or southerly position. It is thought that the alignment of Stonehenge served to point out the most northerly rising point of the Sun, and hence to show when the summer solstice occurred. The later bluestone circles, and various rings of holes which have been found by excavation, may have been used to help calculate the calendar between solstices.

Other circles

There are several hundred sites in Britain where Neolithic or Bronze Age standing stones (also known as *megaliths* or *menhirs*) have been erected, although none is as im-pressive as Stonehenge. At some sites there may be several large circles of stones, but often there are only one or two isolated stones. The purpose of these sites is even more doubtful than that of Stonehenge, but it has been noticed that many of the alignments between menhirs, or between a menhir and the centre of a nearby circle, seem to indicate the point on the horizon which corresponds to the rising or setting of the Sun at one of the solstices. Some of the alignments appear to indicate the furthest north and south setting points of the Moon, whose motion is more complex than that of the Sun, and for many of these lines there is a natural dip or slope on the horizon which could have been used to give a much more accurate position for the rising or setting.

The lunar observations were probably not for calendar use, but for the prediction of eclipses. A solar eclipse occurs when the Moon moves in front of the Sun and blocks off its light, creating a few minutes of darkness, while in a lunar eclipse the Moon enters the shadow of the Earth and fades to a dark red colour, due to some red light being refracted by the Earth's atmosphere into the shadow. These phenomena are particularly impres-sive, and it is possible that the men who built the megaliths were trying to predict them. There is not an eclipse at every new and full Moon, even though the Sun, Moon and Earth are lined up, because the Moon's orbit is in-clined to the Earth's orbit around the Sun, and so it usually appears either above or below the Sun (or the Earth's shadow, in the case of a lunar eclipse). By a careful study of the rising and setting points of the Moon, particularly the most northerly or southerly points, it is possible to predict when an exact line will occur, and hence when eclipses will happen.

At Stonehenge the four *Station Stones* which stand at about the distance of the Aubrey holes may have been used to locate the extreme points of the Moon's motion, and it has been suggested that the 56 Aubrey holes may have been used to predict the years in which there would be an eclipse at one of the solstices. This is because eclipses repeat three times in almost exactly 56 years, and if three stones were spaced equally around the Aubrey holes and moved on a hole every year, a 'danger' year would be forecast whenever one of the stones was in a particular hole. This use of stones to follow the movement of celestial bodies is really an example of an analog computer, and certainly the prediction of eclipses by a people who apparently did not even write is a major intellectual feat.

The study of the alignments at megalithic sites also shows that their builders may have divided the year into sixteen equal 'months', and marked the rising or setting point of the Sun at the beginning of each, while there is good evidence that all the stone circles in Britain were built using the same unit of length, the *megalithic yard*, of 2.72 ft (83 cm). Most of them are an exact number of megalithic yards in diameter; and in cases where the shape is not an accurate circle, it may be that a geometrical construction was made using whole numbers of megalithic yards.

Critics of the astronomical theories point out that eclipse predictions require that observations be made of the Moon when it is rising on a particular night every nine years or so. In Britain's climate, particularly in the Western Isles of Scotland where many stone alignments are to be found, it is often impossible to see the Moon rising because of poor weather conditions. Recent climate research, however, has shown that during the period in question Britain probably enjoyed a slightly warmer climate than at present.

Another criticism is that many of the alignments could have been due to chance. Certainly there are a large number of apparent alignments which have no apparent astronomical significance whatsoever. At this point, the debate rests on the number of alignments so far measured and the accuracy with which the successful lines point out particular phenomena. The subject is still being pursued from this aspect. People often also wonder how a society without writing and apparently without knowledge of the wheel could have carried out such elaborate work. But the people involved are just 160 generations or so away from ourselves, and were essentially modern man. The Mayan civilization of Central America just 500 years ago did not know about the wheel, but they had an even more advanced numerical astronomy. And for a people who could not write, what better way to make a record of movements of astronomical bodies than by setting out markers on the ground in line with bodies themselves?

Above: the monoliths in France near Carnac are intriguing. One theory indicates that they may have been a graph for predicting eclipses.
Below: Callanish in the Hebrides has been called a Scottish Stonehenge.

Roadbuilding and street lighting

The Romans were the first great roadbuilders. Their long straight roads between cities and military camps allowed speedy access for their legions and packhorses. When the Roman Empire fell, the nomadic tribes who were left had no use for these roads which fell into disrepair.

Throughout the Middle Ages, government was enforced by the local baron and so little roadbuilding was done; a road was only the right of way between towns, and most of the trade and heavy loads of merchandise were carried by water or on pack animals, little use being made of wheeled vehicles.

In the 14th century, the local barons lost power to the national monarchs. To enforce their authority, the monarchs required to travel throughout their kingdoms on good main roads, but as the packhorse was still the means of carrying loads on land, no great works of roadbuilding were carried out.

Elizabeth I of England made it one of the duties of the Justices of the Peace to ensure that the local roads were kept in repair. Later, turnpike trusts were set up to maintain the roads and charged tolls on travellers. In 1716, the French king took over responsibility for maintaining his country's roads. Men of ability then began to take an interest in roadbuilding. Trésaguet (1716–1796) in France and later, towards the end of the 18th century, Telford (1757–1834) and McAdam (1756–1836) in Britain applied scientific principles to road construction, so

greatly improving the standard.

Roadbuilding in France and Europe was encouraged by Napoleon, who required good direct roads to move his armies. Several roads over the Alps were then built. These improved roads produced the heyday of the stagecoach, but the advent of the railways in the 1830s stopped this roadbuilding.

Early in the 20th century, the development of motor transport required the improvement of road surfaces using tarmacadam, but World War I delayed a further phase of great roadbuilding. Between the World Wars, the growth of the American car industry produced much roadbuilding, and in Germany, the *Autobahnen* formed the first national express highway system; but construction of the British national motorway network only started in the late 1950s.

Road design

Once the road authority has decided to construct a new major road, then it will employ either its own engineers or a consulting engineer to survey the alternative routes and carry out the road design. Information is required, for each of the possible routes, about the detailed ground levels of the terrain, which can now be obtained by aerial photography which is accurate to 6 inches (152 mm). Details of the types of material for the construction of embankments, and of the geological strata, must be obtained from trial pits and bore holes taken along the

Above: a black-top paver spreading hot asphalt mix which has been tipped into it by a truck which travels ahead. Machines are available which can lay the equivalent of several layers in one pass.

Left: heavy machinery at the early stages of roadbuilding work.

139

line of the route and at bridge sites. The local climatic conditions, such as fog, frost and rain, must also be established. In developed countries, information is required about land values and various environmental factors which may need public enquiries.

From the survey information, the line and level of each of the possible roads will be chosen in accordance with the standards of gradient, sight lines and other factors laid down by the traffic authority. This should minimize the amount of material which has to be excavated and carried to 'fill' the adjacent embankments. It is also important to keep to a minimum the size of the bridges needed to cross railways, rivers and other roads. Taking into account these various factors, the choice of route is made and the design carried out.

The contract
Arrangements are then made to purchase the land on which the road will run. Detailed drawings, specifications and bills of quantities are prepared so that contractors can tender, normally in competition with each other, for the construction of the work. The consulting engineer or highway authority will usually provide a resident engineer and site staff to ensure that the work is carried out by the successful contractor in accordance with the drawings and specifications. Within the requirements of the design, the contractor will be responsible for deciding upon the methods of construction to be used.

Fencing and site clearance
When the construction team moves on to the site, it is first necessary to clear the line of the new road and fence it where animals may stray on to the works. Trees are cut down, stumps and rocks grubbed up by bulldozer or where necessary, blasted out by explosives. It may also be necessary to build temporary haul roads and bridges or fords at the site of the river bridges.

Earthworks drainage
The base of embankments and the slopes of cuttings must be protected from the action of ground water which could cause them to collapse. A primary drainage system is therefore constructed before starting earthworks along the length of the road to cut off the natural ground drainage, and prevent it from entering the works. This is usually done by digging a shallow cut-off ditch with a hydraulic excavator which has a shaped bucket. At the low point of the natural ground, the water flowing in these ditches is taken across the road line in piped or reinforced concrete culverts and allowed to flow away through the existing streams or ditches.

Earthworks
The topsoil is first stripped and stacked ready for spreading on the slopes of cuttings and embankments towards the end of construction. This work is usually done with caterpillar tractors towing box scrapers. The main cutting and embankment work is then started, using rubber-tyred scrapers. These are single or twin-engined machines which have a horizontal blade that can be lowered to cut a slice of earth from the ground and collect this earth in the bowl of the scraper. When the scraper bowl is full—some machines can carry up to 50 cubic yards (38.2 m³)—the blade is raised and the loaded scraper travels to the 'tip' area on the embankment. Particularly in hard digging, it is necessary for the scraper loading operation to be assisted by a pusher bulldozer which pushes the scraper while it is loading to speed up the operation. For certain types of material, such as chalk which may soften in wet weather, or when the excavated material has to be carried

1 cutting

2 embankment

3 geological strata

4 fence

5 primary drainage 'cut-off ditch'

6 piped or reinforced concrete culvert

7 original ground level

8 top soil stripped beneath embankment for re-use on cutting and embankment slopes

9 successive embankment layers

10 road drains in verge and central reserve

11 connection road drain to cut-off ditch

12 verge

13 central reserve

14 carriageway

15 'overbridge' over road

16 'underbridge' under road

17 stream or watercourse

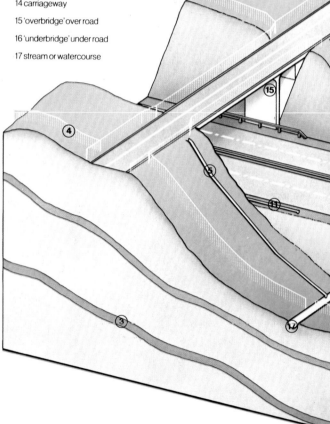

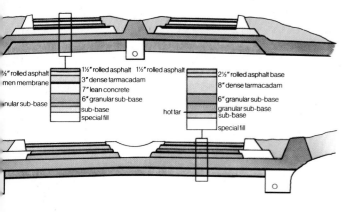

½" rolled asphalt
men membrane

nular sub-base

| 1½" rolled asphalt | 1½" rolled asphalt |
| 3" dense tarmacadam |
| 7" lean concrete |
| 6" granular sub-base |
| sub-base |
| special fill |

2½" rolled asphalt base
8" dense tarmacadam
6" granular sub-base
granular sub-base
sub-base

hot tar

special fill

Top left: this scraper, instead of scraping the earth into a box, has a hydraulically operated conveyer/elevator mechanism which lifts the earth into the box.

Top centre: a scraper and a bulldozer working on a new road in Nigeria. The scraper has two engines, one at each end.

Top right: asphalt rollers have a basic weight which can be increased by means of ballast.

Above: part of a slip-form paving machine, which is laying the sealing strip into a transverse expansion joint cut in the wet concrete.

Left: this diagram includes many of the features which may be used to solve problems in roadbuilding. Bridges such as the one in the drawing often have special surfaces, because they become slippery more easily than the road with earth under it.

for more than two miles (3 km), the excavation may be done using face shovel excavators loading into dump trucks. When rock is encountered, this is first shattered with explosives and then loaded by face shovel.

At the embankment, the earth is spread by the scrapers into a thin layer about 12 inches (305 mm) thick which is levelled by bulldozers and then compacted by caterpillar tractors towing rollers, or by self-propelled rollers which work more quickly. It is vital that the successive embankment layers are properly compacted so that the embankment is stable. The inevitable earth settlement will also be kept to a minimum, thus preventing damage or excessive maintenance to the finished road.

Road drainage
On completion of the earthworks, further shallow drain trenches, about 4 ft (1.2 m) deep, are constructed to keep the top layer of cutting or embankment free from water, which would weaken it. Pipes are laid in these trenches, which are then filled with gravel. In or adjacent to these trenches are laid further pipes to carry the water collected in road gullies from the finished road surface.

Paving
It is necessary to phase the bridge construction periods so that the bridges are completed ahead of the paving

Above: an elevated concrete highway under construction at Chillon, in Switzerland.

Right: crash barriers on a curve in a Greek road above a steep drop. These are designed to give without snapping, so that a car neither bounces back across the road nor plunges over the drop.

operations and the existing roads diverted over or through them. The carriageway [pavement] paving operation then begins by the top layer of earth—the formation—being accurately trimmed to a 2 inch (51 mm) tolerance by scrapers or a grader. A grader is a wheeled machine which has a steel blade mounted horizontally between its four wheels. This blade can be accurately raised, lowered or tilted by the driver to cut a precise surface.

A sub-base layer of gravel or crushed rock, generally 12 inch (305 mm) thick, is then spread over the formation to increase its load bearing capacity. Alternatively, when

Left: a concrete train is almost like an assembly line. This one is laying an unreinforced concrete roadway 22.25ft (7.3m) wide.

Below: an aluminium barrier which carries the idea of flexibility one step further: it is designed with three sections to give way one by one.

New solutions to the problem of a car running out of control have been developed. Tolerable deceleration of the car is provided by means of non-rigid barriers. One method, in use in North America, uses a row of plastic barrels filled with varying amounts of sand; as the car progresses into the barrier at a reducing speed, successive barrels are filled with increasing amounts. This results in a 'controlled' crash even from approach speeds of 60-70 mph. Another method, used in hilly country in Europe, provides an 'arrester' bed of loose gravel, especially at a down-hill bend where a vehicle might leave the road. See picture overleaf.

the earth is suitable and imported materials difficult to obtain, it may be possible to mix a quantity of dry cement into the top layer of earth, which is then damped with water to cause it to harden. This is called soil stabilization.

The final layers of the road are then built, either of concrete or of tar-bitumen black-top materials.

Black-top roads

The bitumen and stone are heated and mixed together in a site mixing plant, and brought hot, by truck, to the laying point. The material is then spread evenly to a thickness of 2.5 inch (64 mm) and to an accurate level.

This layer is compacted by road rollers to give a firm surface. The accuracy of each successive layer until the final wearing course (usually of asphalt) provides the accuracy of the finished road surface. The total black-top thickness will be about 12 inch (305 mm). To improve the skid resistance of the road, bitumen coated stone chippings are spread over the top surface and rolled into it while it is still hot.

Concrete roads

If the final surface is to be concrete, then this will consist of a concrete slab approximately 10 inch (254 mm) thick.

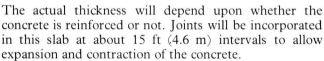

The actual thickness will depend upon whether the concrete is reinforced or not. Joints will be incorporated in this slab at about 15 ft (4.6 m) intervals to allow expansion and contraction of the concrete.

Conventionally, the concrete is laid between temporary steel road forms, which support the edge of the concrete slab, by a concrete train. This consists of a series of machines which run on rails supported on the road forms. The forms and thus the rail are accurately laid to level well ahead of the train and provide the level control for the finished road surface. The first machine in the train is a placer spreader which puts the concrete, transported by truck from the concrete mixing plant, between the road forms. The concrete is then compacted and trimmed to true level by successive machines. To provide a skid resistant surface, the wet concrete is then lightly brushed or otherwise grooved to a shallow depth. Pockets for catseyes are also formed in the wet concrete at this stage.

Recently slip form machines have been developed, and by using these machines to form the concrete slab, it is possible to eliminate the lengthy process of accurately laying out road forms. These slip form pavers incorporate travelling side forms within the body of the machine. The degree of vibration compacting the concrete is much greater than with the conventional train so that after the moving forms—approximately 15 ft (5 m) long—have slipped past, the fresh concrete is able to stand up without further support. The surface level of the finished concrete is formed by the same machine which is controlled, both for level and direction, by means of electronic or hydraulic sensor controls which follow string lines placed at each side of the machine along the pavement. With this paver, it is possible to achieve up to 6 ft (1.8 m)

per minute.

Road finishings

Once the carriageways are completed, it is possible to fill up the verges with earth and then topsoil and seed the cutting and embankment slopes. Safety barriers can be erected in the central reserve between the lanes or in the edge verges where the height of embankment may justify them. Catseyes (reflectors) may be laid on the lanes to indicate the traffic lanes and slip road entrances, together with white lines. These white lines are composed of small spherical glass beads contained in a soft plastic paint which is sprayed on to the road surface giving a reflective effect.

Street lighting and signposts are erected at the intersections and roundabouts where traffic joins or leaves the new road. In towns and foggy areas, overhead lighting is often installed for safety reasons. Emergency telephone cables connected to telephones placed at intervals along the route are frequently installed in the verges. In urban areas it is possible to reduce the level of noise from traffic on the road by erecting light screens about 8 ft (2.4 m) high continuously along, and close to, the edges of the road.

Road safety

In an effort to reduce the numbers of traffic accidents, over the last thirty years the field of road safety has expanded from empirical attempts with legislative moves like urban speed limits and propaganda campaigns into a comprehensive study of the whole environmental, vehicle, road user complex of modern traffic.

Studies of accidents show that the environment and road users themselves are by far the most important factors in the occurrence of accidents. Vehicles themselves

Opposite page, far left: testing a breakaway lamp standard. In this picture the mast has just broken off.
Top: this articulated truck has been stopped from a speed of 25 mph (40 kph) by a bed of graded gravel.
Below: in Cumbernauld New Town, Scotland, pedestrians and cars are kept on different levels.
This page, left: a non-skid surface being applied in the Strand, London. The system, called Shellgrip, uses calcined bauxite abrasive embedded in a resin-based compound. It reduces the greasy effect of roads when they are slightly wet in cities with oily air.
Below: high-pressure sodium lighting shows true colours. Low-pressure sodium lights give everything a yellow cast, making objects harder to distinguish.

cause accidents less often, but have a great influence on the resultant injuries.

Almost every aspect of road design has an effect on accidents. In new towns it is possible with modern layouts to design environments which are very much safer than the rest of the country. For example, the town of Cumbernauld in Scotland has one third of the national average of pedestrian accidents per head of population, because of the segregation of pedestrians from cars, with the provision of subways, bridges and special pedestrian routes. Similarly motorways, because of restricted access and the absence of pedestrians, parked vehicles and turns across oncoming traffic, have about half the number of accidents per vehicle mile travelled than other classes of road. But since most town and country roads have been developed before the invention of the motor car, road safety is mainly concerned with modifying what is already in existence.

The basic aim of all road alignments, junction design and road signs is to simplify the decisions which have to be taken by drivers and provide enough time for those decisions to be made correctly. Thus, improving the alignments at bends with bad accident records gives up to 80% reduction in the accidents occurring at those sites. At crossroads the provision of roundabouts, traffic signals, or the staggering of the minor roads similarly reduce accidents by about one half.

Road surfaces influence the occurrence of accidents profoundly, and in recent years special materials have been developed for sites where heavy braking is frequent. One such surface is a calcined bauxite in a resin base; this provides a very high skid resistance even when wet, and does not polish significantly with use. It has reduced

accidents by 31% at 41 experimental sites in London. At present the cost limits its more widespread application although, in common with most other road safety measures which modify the environment, the returns obtained are substantial. In Britain in 1972 the average injury-producing accident cost £1300. The criterion for the implementation of a remedial measure is that between 50% and 500% of the cost of accidents at a given site should be recovered in the first year. There is no other area of public expenditure where the returns on money invested are so great.

The advent of high speed divided roads has created an additional problem for the road surface in wet weather. At speeds over 50 mph (80 km/h) if water collects on the road, a wedge of water intrudes between the moving tyre and the surface, and so reduces the effective area of contact. If the water becomes established completely beneath the tyre, then all direct contact is lost and 'aquaplaning' occurs. Modern tyres have been developed with tread patterns designed for draining, but a more effective solution is to use a road surface which is porous, thus allowing rain to drain away into the surface instead of having to let it run off into gutters at the side of the road.

Lighting

One of the first references to any form of street lighting was in 1405 when Aldermen of the City of London were made responsible for ensuring that every house adjoining the highway provided a lit candle in a lantern from dusk until 9 p.m. Since then the technology of street lighting has made great and significant advances, particularly in view of the recent development of new light sources and the increased density of fast moving traffic.

Street lighting is often referred to as 'road lighting' or 'public lighting' which includes the lighting of pedestrian walkways and subways. The main object of street lighting is to ensure that ever increasing numbers of vehicles and pedestrian traffic can use the roads safely during the hours of darkness. An important factor in safety is the reduction of fatigue in the driver and the removal of causes of annoyance. About 35% of all accidents occur during darkness and the installation of street lighting on an unlit road can reduce night accidents by between 30 and 50%.

The problems of designing and commissioning a public lighting installation are more complex than would appear at first sight. To a large degree it is still an art as well as a science, in that it demands a large measure of sensible and personal judgement.

Seeing by what is here called direct vision—that is, with all objects well illuminated—is costly as very high levels of lighting would be needed. In practice the basic theory governing street lighting is that the road surface and surroundings should appear light in order that vehicles and objects appear in silhouette. The road, and if possible the surroundings should be lit brightly and evenly so objects can be easily observed. Reflection of light from the road and pavements is therefore of prime importance, so surfacing materials having good reflective properties should be used.

If a bare lamp is suspended over a road and is shielded by a dense shade to such an extent that an observer further along the road cannot see the lamp, then the observer will only see a bright patch of light on the road surface. On this basis lamps would need to be set very close together to allow the patches to merge so that there would be uniform brightness.

In practice the lamps are not directly shielded, but are contained in enclosed weatherproof *lanterns* from which light is not wasted upwards but is directed at certain well-defined angles by precisely designed shaped reflectors and refractors.

There are two main types of lantern used for street lighting and both are designed to produce T shaped patches on the road for the distant observer. One is called a cut-off lantern, and the other a semi cut-off. Cut-off lanterns have a beam angle of about 65° to the vertical, and the semi cut-off lantern has a beam angle of about 75°. The total angle as seen from the lantern is twice these values.

Semi cut-off lanterns are used where there are smooth road surfaces as these give good reflection and extend the tail to the T shaped light patch. The cut-off lanterns give better glare control, and are used for rough road surfaces which give poor reflection. Thus the tail of the T is reduced and this requires a closer spacing between lighting points.

The most favoured lantern arrangement over the road is the staggered system, as this provides lighting uniformity with reasonable efficiency. In this system lanterns are placed alternately on one side and then the other. For wider roads lanterns are positioned opposite each other and on four or six lane motorways two lanterns on a single column installed in the central reservation, if it is not too wide, is the most economical and favoured system.

High intensity lighting from masts up to 150 ft (46 m) high can be used at large roundabouts and multi-level junctions to reduce the number of conventional lamp columns, which can be road hazards. With high mast lighting we see by direct vision. In tunnels, lighting is provided to a relatively high level to enable a driver to make immediate decisions by day or night.

For efficient and economical street lighting low pressure

sodium discharge tubes are ideal, and for areas where good colour rendering is required, such as city centres, high pressure sodium lamps are now being used. Until recently, colour corrected mercury lamps were used in residential and city areas because they give a somewhat whiter light than sodium. They are, however, expensive to run and are now rarely used in new installations.

The present and future needs to conserve energy costs will escalate the replacement of tungsten and high wattage mercury vapour lamps by the highly efficient sodium light sources.

The first controller to put street lighting on and off was the 'lamp lighter', who went around lighting gas lamps, and since then there have been many methods of control to switch lights individually, in groups, or in large numbers.

For many years the most reliable and economic system has been the solar dial time switch. This is a clock driven by a synchronous electric motor. The solar dial is itself compensating and allows automatically for the daily variation of switching on and off times which relate directly to sunrise and sunset.

The disadvantages of the solar dial time switch is that its automatic switching of street lighting is based on average daylight conditions, and it needs re-setting after every power cut. There is a very strong argument for a switching device controlled by ambient daylighting conditions, for example to switch lights on when there is fog or when it is dusk early owing to an overcast sky.

The modern photo-electric cell fulfils this requirement and is being used extensively for new street lighting. A photo-electric cell can be seen generally on the top of a lantern controlling an individual lamp, although it is quite often used wired to a relay which in turn switches large numbers of lights.

Other aspects

Speed limits represent only one aspect of traffic management which has benefits for road safety. In urban areas the reduction of congestion, improved flows and the reduction in accidents are all complementary aims of good traffic control. Vehicle actuated traffic signals at junctions for example allow changes in daily flow patterns to be accommodated, and together perhaps with 'all red phases', allow conflicts at the junctions to be reduced. The linking of traffic signals to a computer allows maximum flows and minimum delays to be obtained along an arterial road or over the whole of a city centre.

Pedestrian fatalities in most European countries represent almost one third of the total number of traffic deaths. Various procedures operate aimed at segregating in time or space the car from the pedestrian. Designated crossings of the zebra variety where the pedestrian has priority are common, but pedestrian operated light signals are being used increasingly in Europe. Underpasses and bridges are safe but expensive solutions, and studies have shown that pedestrians are very willing to take risks if the safer alternative route is appreciably longer.

As far as the environmental aspects of road safety are concerned, solutions are available to almost every problem, but the cost of the desired changes restricts their introduction. Road user behaviour, however, is more difficult to control, as there is very little connection established between training and codes of conduct on the one hand, and actual accident involvement on the other. Road safety therefore is mainly an engineering problem, where the system is designed and operated in a manner which can accommodate the considerable range of human capabilities and behaviour. Attempts to modify human behaviour on the whole are unsuccessful and good traffic control and road safety measures now recognize this fact. 147

Rubbish disposal

Although drainage systems for sewage have been in existence for at least 5000 years—there were sewers at Nineveh in 3000 BC—organized procedures for the disposal of refuse or solid wastes are a comparatively modern invention. When the stacks of rubbish produced by primitive communities became too offensive to bear, there was always the simple expedient of moving on to another site. The increasing size of urban societies has exacerbated the unpleasant problem of dealing with the cast-offs of daily living, however, and the recent proliferation of non-organic containers and packaging materials has created new difficulties. In 1970 the average British household generated 35 lb (16 kg) or rubbish per week, American households produced an astonishing 53 lb (24 kg), and in 1971 the disposal of solid wastes cost the American people $4,500,000,000. About 1 square foot of ground was once required to dispose of 1 cubic foot of waste material, but much modern rubbish is not biodegradable (naturally decomposing), and this factor, in combination with the almost universal shortage of land, has produced a huge problem.

The technology of waste disposal has a relatively short history. In Britain there were no requirements for the disposal of refuse until the Improvement Acts of the 18th century. Then in 1875 the Public Health Act was passed, and its stipulations form the basis of most of today's refuse removal and disposal regulations. Prior to the legislation which has been enacted in most developed nations, waste disposal was left to the individual householder, who moved his rubbish when and where he could.

Removal

Modern waste disposal presents three problems: removal, storage and final disposal. The dustman continues for the average citizen to represent the end point of the waste disposal problem. Labour for collection and transportation makes up the most of the expenditure on waste disposal, and, in America, nearly 80% of waste disposal expense arises from collection and transfer of refuse. From the householder's point of view, probably the most significant event in the history of waste disposal has been the requirement in some communities of standard covered containers or tied plastic bags.

The first purpose-built collection vehicle made its appearance in 1922. Collecting vehicles now feature two-way tipping, compaction, side-loading and other hydraulically operated features.

In some communities the dustman has become obsolete. One Swedish residential area, comprising 140 units, is served by a pneumatic system consisting of refuse chutes and an underground main extending as far as one and a half miles. The air in the main is sucked out by turbo-exhausters and the resulting vacuum causes valve plates on the chutes to open and air to be sucked in from the storage areas at the tops of the chutes. Refuse is drawn into the underground main and carried at up to 50 mph (80 km/h) to a silo. The refuse is dumped into the top of the silo, while the air is drawn off from the main by the exhauster for filtering before it is returned to the atmosphere. Other advances include the electric waste baskets which have been marketed in the USA. They destroy, shred and compact the bulk of paper ingested. Miniatures of large-scale compacters are also sold to private citizens; some are capable of reducing the volume of rubbish by 90%, and even of wrapping the final product in a cellophane bag. Private waste disposers fitted under the kitchen sink can also minimize the amount of distasteful organic garbage to be collected by the dustman. Powered by an electric motor, it is connected to the outlet pipe and consists of grinders which are driven at speed to produce a suspension of organic solids that is fed to standard sewage systems.

The increasing concern of environmentalists and conservationists has raised the level of public consciousness of what happens to rubbish once it has left the home. Similarly, local authorities and individual communities have had to face up to the implications of land shortages for waste disposal. The vast majority of rubbish in the UK is disposed of by tipping, and in the USA sanitary landfill is employed. The old public dump is inefficient, unsightly, uneconomical and dangerous. By 1916, in England, controlled tipping was being tried. This consists of laying wastes in shallow layers and covering successive layers with soil to compact them. The soil layers are usually less than six feet (2 m) in depth to permit aeration and natural decomposition, and care is taken that waste has been previously crushed and compacted so that the eventual collapse of large objects will not result in an uneven surface. The system is used for land reclamation; many tip sites have been turned into parks or even, in the USA, into artificial ski slopes. Nearly 90% of wastes in Britain are dealt with in this way, but shortages of space have meant that the waste must first be treated and reduced in volume.

There are two main techniques for treating 'raw' rubbish: pulverization and incineration.

Pulverization

Pulverization is a mechanical means whereby rubbish is crushed, screened, shredded and so on to decrease its volume. The chief methods are wet and dry pulverization. Wet pulverization utilizes the power of water, thereby reducing the need for more expensive sources of energy, but its use is more or less limited to on-site tipping because of weight considerations and the problems of waste transfer. A wet pulverizer typically consists of horizontally rotating drums. Water is sprayed into the drum as waste enters and pulverization occurs because of the pounding effect of deflector plates, baffles and rotation. Inner and outer drums are often used, with a separating screen, so that the pulverized waste in the central drum is forced through the screen into the outer drum.

One wet pulverizer, 8 feet (2.44 m) in diameter and 29 feet (8.84 m) long, will pulverize 60 to 70 tons of

The storage bunker of this municipal incinerator is large enough to hold a 24-hour collection. The rubbish is gravity fed to a combustion chamber.

refuse in a 7 hour work day. The Dano Egsetor Pulverizer moves at about 12 rpm, is powered by a 60 hp motor on an external gear ring and is completely covered by sheet steel housing. The inner chamber is formed by manganese steel bars in a grid, and the outer chamber is made of manganese steel plates. Refuse is fed into the inner drum, moistened by water to weaken cardboard and paper and tumbled around until it has disintegrated sufficiently to pass through the steel bars. Granulation continues in the outer chamber.

Dry pulverizers use shredders, hammers and rotors to reduce the volume of refuse. In one model the refuse enters at the top, is drawn downwards by contrarotating motors and shredded by teeth at the base of the mill. Most are of the hammer mill type with a closed casing, hammers attached to discs or fitted directly to rotor shafts, and shredding teeth or a grid. The Gondard Refuse Reducer incorporates a separating device as well and thus eliminates some of the maintenance problems which beset waste disposal units. Refuse is fed into the machine from a direction opposite to the rotor; upswinging hammers hurl uncrushable material up into a hopper. Dry pulverizers crush glass to a fine powder, which the wet models do not, and greater control of particle size is permitted. Some machines are capable of more than 2000 rpm and can pulverize as much as 70 tons of refuse an hour.

Incineration

The most effective way to decrease the volume of refuse is incineration, although conservationists are becoming increasingly concerned about the advisability of burning materials which may soon be in short supply. They are similarly concerned about the energy requirements of systems used to destroy organic and inorganic matter, so that some authorities have experimented with using the energy produced by the incinerating process to recover expenses. One American scheme, in New Haven, Connecticut, involves three separate housing projects and a sort of perpetual motion in the transference of energy. One building houses an incinerator and two others each have a compacter and a wet pulverizer-shredder. Sink waste disposers have been installed in each flat, and tests have been tried in which the power from the incinerator is used to generate electricity, which in turn drives a vacuum pipeline leading from the individual disposers to the incinerator.

Incineration typically consists of three stages: drying, combustion and a final stage to render the waste burntout. Drying can be achieved through the heat from refuse and fuel which are already burning; by thermal radiation, convection, pre-heated air and waste gases. Automatic grates on which the refuse is placed at the top end are often sloped so that hot gases from the main combustion area flow over new material. The main combustion is often assisted by highly combustible fuel. Temperatures at this stage need to be over 750°C (1382°F), but at temperatures in excess of 1000°C (1832°F) ashes will fuse. A clinker (the end product of incineration) which is essentially free of organic material depends on the combustion of slower burning materials, so that wastes transported through the main combustion stage must have an opportunity to burn out. Normal requirements are that clinker should contain no more than 0.3% putrescible material, and free carbon should constitute less than 5.0% of the residue.

Refinements on the technology of refuse incineration have concentrated on automation, improving the grate and control over incineration rates. The Torrax system is a different sort of process. Here refuse is fed into the top of a gasifier and settles toward the high temperature zone at the bottom. Very hot gases from a hot blast heater rise through the refuse, causing some materials to burn before reaching the bottom, and non-combustible materials such as glass and metals melt and flow into a water-filled chamber. Pyrolosis, a gasification process involving the application of heat in the absence of air, produces recoverable and potentially valuable by-products, such as carbon monoxide, methane, hydrogen, char, oils, tar and metals. Refuse is fed into a chamber and heated to temperatures in excess of 600°C and less than 1000°C (1112–1832°F). The process produces a 10 : 1 reduction in volume.

Pollution control

In less complicated times, incineration was widely employed by individual households and local authorities without special concern for pollution. The unpleasant effects of soot and dust were accepted, but the development of many noxious and potentially dangerous chemicals has resulted in fairly strict controls as to the content of gases and smoke sent into the atmosphere by incinerators. Various gas cleaners are used to ensure that levels of dirt and ash in flue gases are acceptable, and some ingenious devices have been developed. Perhaps the simplest technique is the settling chamber of greater cross section than the flue, which because of expansion of gases causes a decrease in velocity so that heavier particles

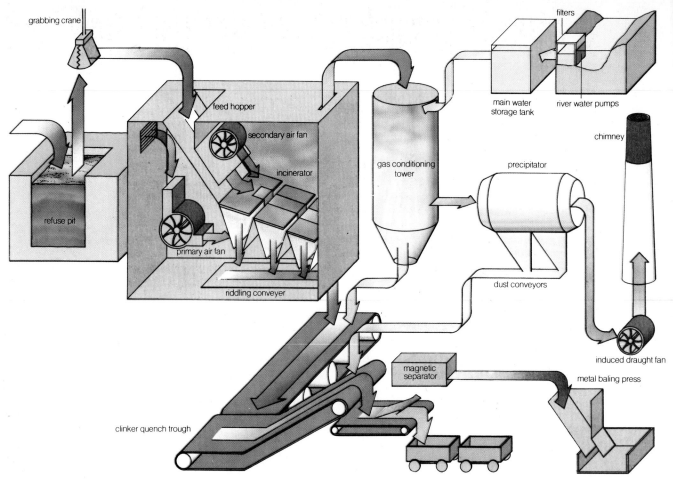

grabbing crane

feed hopper

secondary air fan

incinerator

refuse pit

primary air fan

riddling conveyer

clinker quench trough

filters

main water
storage tank

river water pumps

chimney

gas conditioning
tower

precipitator

dust conveyors

induced draught fan

magnetic
separator

metal baling press

fall into a water trough. Another inertial separator is the cyclone. Here the dirty gas from the incinerator enters a cylindrical cell tangentially to create a vortex. The resulting centrifugal force causes dust particles to move outward to the cell wall. The stream of gas is directed downwards so that debris moves in a spiral toward the bottom of the cell where it is collected. The gas itself, not as subject to centrifugal forces, reverses its direction and discharges from the top.

Gas scrubbers are also used to clean waste gases from incinerators. Dirty gas passes through a spray of water where dust particles become entrained. Still more effective is the impingement technique of fabric filters, in which dirty gases pass through various kinds of fabrics. Systems incorporating fabric filters are expensive, however, since pores quickly become clogged and the filters must be replaced frequently.

The technique most used in purifying gases produced by the incineration of municipal wastes is electrical precipitation. The principle of precipitation of dust particles is that charged particles in an electric field will move towards the electrode of opposite polarity. The precipitator consists of alternate rows of high voltage discharge, ionizing wires and earthed 'collecting plates'. Direct current is applied between the wires and the plates (positive electrodes) so that negatively charged ions collide with dust particles in the gas. The particles obtain a negative charge, therefore, and are attracted towards the collecting plates. The clean gas continues on its way while dust and ashes are left adhering to the plates.

Composting

Composting has perhaps the longest history as an organized process of waste disposal with a view to return-

In this typical system, the rubbish is incinerated and the hot combustion gases cooled by a fine water spray; then electric precipitators remove the dust. Rubbish which won't burn is hauled away.

ing useful elements of refuse to the land. The Romans are believed to have dealt with their organic wastes in this way; early methods included methodical layering and turning of waste to aerate it and to assist biological activity.

The chemistry of composting is extremely complex, and mechanical techniques have evolved a high degree of sophistication. To be suitable for composting, wastes must contain sufficient levels of sugars, proteins, fats, hemi-celluloses (polysaccharides of 50 to 150 sugar units), celluloses (polysaccharides of 1000 to 10,000 sugar units) and lignin (tough cell wall material). Composting of municipal wastes is not practised widely in Britain, because the final product is often unsuitable, since municipal wastes contain significant amounts of metals and trace elements.

Mechanical composting concentrates on ways to improve the rate of decay by artificially creating optimum conditions. Careful control is exercised over concentrations of nutrients for micro-organisms, particle size, homogeneity, moisture content, temperature, agitation and aeration. Horizontally rotating drums and multi-deck plants where new compost material works its way to different levels and finishes on the bottom have been tried, but the most sophisticated techniques are used in the multi-deck silo digesters. These are enclosed so that greater control is possible, and they offer the best hope for future commercial composting.

In countries where water is plentiful and cheap, people often tend to abuse it, using much more than they really need. It is used for drinking, bathing, washing clothes and cars, for a thousand and one different purposes in industry and finally, after it has been used, it is thrown away down a toilet, a waste pipe or some convenient drain. In industrialized countries, each person on average uses about 50 gallons (227 litres) of water per day, and rivers and streams would be in an appalling state if all that water were allowed to reach them in a dirty condition, bringing with it a load of filth and waste materials.

The need for something to be done about the cleaning up of the population's waste waters, or sewage as it is generally called, became apparent about the time of the Industrial Revolution. Initially the ability of organisms, normally present in soil, to remove or stabilize polluting matter was used extensively. For example, organic matter of animal origin contains nitrogen, often tied up in a complex form, but the soil organisms are able to break down the complexes and incorporate the nitrogen into

the soil itself. So, by irrigating sewage on to land a certain amount of purification of the water was effected. But any good gardener knows that continuous fertilizing of land will upset nature's balance in the soil unless suitable crops are grown as part of the cycle—and so the term 'sewage farm' became part of the language. As a rough rule of thumb guide, at least 100 acres (40.1 hectares) of land would be required for each one million gallons (4.5 million litres) per day of sewage to be dealt with (which would be the sewage from a population of about 20,000).

Precipitation

In due course, other factors began to appear and to affect the problem. In Britain a Royal Commission was set up in 1898 to investigate the whole question of sewage disposal, and in its eighth report it made certain recommendations governing the standards of impurity of sewage effluent entering a stream or river—those same standards are in many cases still quoted and applied today. Because the average sewage farm was not able to produce effluents complying with the Royal Commission Standards, and because increases in population inevitably required more and more acres under irrigation on the farms, efforts were made to concentrate the process so that less land would be required.

Chemical precipitation was introduced, whereby a small dosage of a salt of either iron or aluminium was introduced into the sewage, followed by the addition of a suspension of lime in water to raise the pH value. The

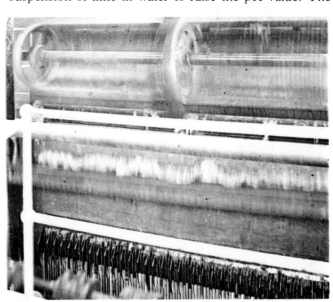

Above: a fine bar screen which removes solids from sewage; a revolving brush cleans a screen-clearing rake. The solids removed are burned or broken up and returned to the flow for further processing.

Left: the grit-collecting trough of a constant velocity channel. Grit might damage the pumps unless it settles out.

Above left: a primary sedimentation tank, with the floor cleaning blade lowered. The travelling bridge drags the blade along to clear the tank of sediment after it has been drained.

Left: a biological filter. Settled sewage is distributed onto the porous material by a rotating sprinkler arm, and colonies of organisms oxidize the impurities as the sewage trickles through.

Above: diffusion domes in an aeration tank blow air up

pH value of a solution is a measure of its acidity; a neutral solution has a pH value of 7, acids have values between 0 and 7, and alkalis between 7 and 14. This caused the hydroxide of the metal used to be precipitated, and as the treated sewage was then passed through large tanks, the precipitate would settle to the bottom of the tank, taking with it a considerable amount of the impurities present. While the resulting effluent was much more amenable to further treatment on land, the process gave rise to large volumes of offensive smelling sludge which had to be removed from the tanks at intervals of one or two days.

Biological filter

Probably the most significant breakthrough, however, was the development of the biological filter. This consists of a basin with side walls between 5 and 6 feet (1.5 to 1.8 metres) deep and a slightly sloping floor to permit free drainage. The basin is filled with some durable material, preferably with pores to give a large surface area per unit volume (clinker, hard coke and blast-furnace slag are all first class). Settled sewage, when sprayed evenly on to the surface of this medium, leads to the establishment in the filters of colonies of the necessary biological organisms to oxidize the impurities in the sewage to stable substances. For example, carbonaceous matter is oxidized to carbon dioxide, and nitrogen first to nitrite and finally nitrate. The whole idea is to create an environment in which the organisms can live, do their job and propagate, and once established, a filter bed will continue to function literally for ever unless it suffers physical breakdown or damage from external causes.

Activated sludge

This particular stage of purification has been further concentrated by the discovery by Ardern and Lockett in 1916 of the Activated Sludge Process. In this process the settled sewage is mixed with activated sludge, which is a suspension of coagulated matter on which live the organisms necessary for the biological purification of the sewage. The mixture (mixed liquor) is passed through specially designed tanks where air is supplied, either by blowing compressed air through fine diffusers or by mechanical means, using rotors which cause air to be taken in from the atmosphere. The sewage and organisms are kept in intimate contact for about 9 or 10 hours, during which time the impurities are biologically oxidized and stabilized. The activated sludge is allowed to settle out in tanks and the clear effluent passes forward to the river.

Sewage itself consists of dirty water containing not only impurities in solution, but also in the colloidal state, in the form of fine sediment and as bulky solid matter, originating as it does from washbasins, toilets, bathrooms, kitchens, and a whole range of industrial processes. The water simply acts as a vehicle for the polluting matter and is not itself changed, so that in theory it ought to be possible to remove the polluting matter and restore the water to its original pure condition; in practice over 90% purification is frequently achieved.

Modern treatment

In a typical modern treatment works, the sewage is first passed through bar screens to remove bulky solids which are either incinerated or are macerated (softened) and

through a mixture when the tank is full. Oxygen is necessary for the biological purification.

Above right: the aeration tanks in use. The mixture of sewage and activated sludge is agitated by the air, which ensures thorough mixing and aeration.

Right: rotating scraper blades clean the floor of this circular final sedimentation tank.

put back into the sewage flow. The next step is to remove grit, which might otherwise damage pumps and other equipment, by passing the sewage through constant velocity channels which are of parabolic cross-section to allow the grit to settle out while the organic matter is carried forward. The grit is removed and disposed of on land.

The sewage is next passed through primary sedimentation tanks where the finer sediment separates out and settles to the bottom as sludge, from where it is removed possibly twice daily, and thickened up a little in consolidation tanks before being pumped to mechanical dewatering plant. Here it is conditioned with chemicals and pressed to form a partly dried cake which is finally disposed of as a fertilizer on farm land. Alternatively the consolidated sludge may undergo heated, anaerobic (in the absence of free oxygen) digestion in large digesters, where the organisms break down the fatty matter and some organic material, thus changing the nature of the sludge and rendering it inoffensive and ready to be sprayed on land as a fertilizer. A by-product of digestion is a gas rich in methane, which is collected and used for heating and power production purposes.

The settled sewage passing over the weirs of the sedimentation tanks is subjected to biological oxidation either in biological filters or by the activated sludge process, both of which have been described. In either case the impurities in solution are oxidized and stabilized and the effluent then passes through final settling tanks to remove any sediment which may have been formed. This sediment, which contains the organisms that break down

the sewage, is returned to the system to keep the biological processes going. At this stage, the effluent is often fit to discharge to the river, but where very high standards are to be complied with, this effluent would be further treated to remove the last traces of suspended matter by passing through sand filters or through microstrainers, where the apertures in the mesh can be as small as 23 microns (0.0009 inch). The effluent is now probably cleaner than the river water into which it discharges.

Other systems

One of the simplest forms of sewage disposal, used mainly in remote rural areas for the disposal of domestic and agricultural sewage, is the cesspool. This is a tank or pit, watertight to prevent leakage into watercourses, into which the sewage is drained. It is emptied two or three times per month, and the effluent is used as manure, often after chemical deodorization and the addition of lime or bleaching powder to kill the unwanted organisms.

A more effective method for handling the sewage from domestic sources is the septic tank. The sewage takes about 16 to 24 hours to pass through the tank, where it is decomposed by anaerobic bacteria. The sludge settles at the bottom of the tank, which is emptied when it is $\frac{1}{3}$ to $\frac{1}{2}$ full of sludge, removing about $\frac{2}{3}$ of the sludge and leaving the rest to maintain the bacterial activity in the tank. It was originally thought that this process would destroy pathogenic (disease-producing) bacteria, but this is now known to be untrue.

In coastal areas, sewage is often discharged directly into the sea without any form of treatment, but increasing use is now being made of methods of diluting or disinfect- 153 ing the sewage to reduce the pollution of coastal waters. One simple method is the use of comminutors, machines which trap the suspended solids in the sewage and cut them into small pieces to speed up dispersal and decomposition. The effluent is thoroughly diluted at the discharge point to prevent excessive local pollution.

A more advanced system is the electrolytic process, in which sea-water is electrolyzed and mixed with the sewage before it is discharged into the sea. During the electrolysis, hypochlorite compounds which disinfect the sewage are formed at the anodes, and magnesium hydroxide formed at the cathodes acts as a floc to precipitate the suspended solids into a sludge which can then be removed and disposed of. The amount of electrolyzed sea-water needed varies from about $\frac{1}{60}$ to $\frac{1}{20}$ of the volume of sewage to be processed. This system is already successfully in use at several pilot plants.

Water supply
Man has always depended on natural sources such as rivers, lakes, springs and artesian basins for his water supplies. As society has become more industrialized with large urban populations, the control of water supplies has naturally needed more careful supervision. Demand often exceeds supplies, so storage in the form of reservoirs or manmade lakes is necessary. In addition, pollutants, not only from industrial effluents but also agricultural sources such as fertilizers, have increased, necessitating the monitoring and purification of raw water supplies, although it would be unfair to claim that pollution of rivers is wholly a 20th century phenomenon. Even in the 19th century it was recognized as a problem: London's River Thames was so polluted that the smell wafting through the windows of the Houses of Parliament prompted the members to pass the first Act to combat pollution, in 1856.

In countries with a high standard of living, man readily 'consumes' about 50 gallons (227 litres) of water per day, although his basic bodily intake is only about two pints (1.1 litres).

Water quality
Municipal water supplies must be drinkable, and various simple tests are performed to monitor its quality. These include tests for taste, odour, colour, turbidity (caused by suspended material such as clay particles), dissolved solids, pH (acidity or alkalinity), and bacteriological and biological contaminants.

Taste and odour may be caused by the presence of hydrogen sulphide given off by decomposing organic matter, or algae, or even chemicals such as phenols, of which only a few parts per million in association with traces of chlorine used for purification is sufficient to produce noticeable taints. Many serious diseases are waterborne, for example, typhoid, cholera, dysentery and infectious hepatitis.

Water works aim to deliver neutral water, that is at pH 7, or a slightly higher pH, which means it is alkaline. This is preferable to acidity, which would cause more harm. Water sources, however, may be naturally acidic from dissolved carbon dioxide forming weak carbonic acid or sulphurous acid from dissolved sulphur dioxide, a common atmospheric pollutant in industrialized areas. Some purification processes may also leave the water acidic.

The test for assessing bacteriological quality is the coliform count. This is called after *E. coli*, nonpathogenic bacteria present naturally in man's intestines and excreted

daily in their millions, and if present in a water sample would be an indication of other possible harmful bacterial contamination. The standard is simple: less than one coliform organism per 100 ml of water—in other words, none. There is also a biological or algal count which measures the amount of microscopic plant and animal life present in a given water sample, other than bacteria.

Another test concerned with water quality is the BOD or biochemical oxygen demand, which measures the amount of oxygen absorbed by a sample of water, maintained at 20°C (68°F), in five days. Fresh clear water is normally full of oxygen and the balance is maintained by the biochemcal functions of the aquatic plant and animal life. If, however, more oxygen is used up through the presence of pollutants faster than it can be recreated or reabsorbed, then the water becomes impure.

Chemical analyses may also be carried out to determine the iron, manganese, lead, nitrogen (the presence of which indicates organic pollution), and calcium carbonate (indicates water hardness) content of the water.

Contamination from radioactive fall-out or waste may also need to be monitored regularly and the WHO (World Health Organization) give the maximum tolerances of one $\mu\mu$curie per litre of water for α-emitters, and 10 $\mu\mu$curie per litre for β-emitters where a population is

likely to be exposed to this form of contamination for a prolonged period.

Water treatment

Both sedimentation, that is, allowing any suspended matter to fall to the bottom, and filtration through sand and stones have been known since ancient times, and large sand filters were first used in London by the Chelsea Water Company to filter Thames river water in 1828. Nowadays there is a wide range of processes available for water treatment, but those commonly employed are sedimentation, flocculation and coagulation, filtration, aeration and sterilization.

Simply storing water in reservoirs assists purification by allowing the larger particles to settle out, by buffering water quality and by letting sunlight hasten the destruction of bacteria. One disadvantage, however, is that reservoirs have a tendency to develop algal blooms, a condition which has been exacerbated by the widespread use of nitrogen fertilizers, the run-off from agricultural lands having entered water sources. Copper sulphate is therefore added to the reservoir to kill off the algae.

The first stage in water treatment usually includes some form of coarse screening at water intakes to remove leaves and so on. From then on there may be variations, for example, chlorine may be added twice, as a pre-treatment and a post-treatment. A typical sequence may be as follows.

Water enters the works via a take-off tower enabling it to be drawn from selected levels, is passed through activated carbon to remove taste and odour that may be present, and receives its first dosage of chlorine for bacteriological control. In modern works it will then pass to the microstrainer, which is a very fine stainless steel mesh cylinder capable of trapping algae and similar sized material, before the water undergoes flocculation and coagulation.

Some particles present in water are so fine that they have to be trapped or coagulated by a sort of chemical 'blanket'. Such particles include colloids which cause turbidity and coloration, and bacteria. Chemicals commonly added as coagulants may be aluminium sulphate, ferric sulphate, or ferric chloride. To assist coagulation, starch is often added.

The chemicals form ions with strong positive charges which attract the fine particles and some bacteria because they bear negative charges. For example, when aluminium sulphate is added to the conical sedimentation tanks containing water which has been rendered alkaline, it reacts chemically with the alkali to form aluminium hydroxide, which is a flocculant precipitate. This forms a

155

'blanket' about half way down the tank; water enters the tank from below and passes slowly through the blanket, which acts as a filter, bringing the positive and negative ions into close contact so that they coagulate.

The clear water passes to the filters, which may be slow or rapid. One type of rapid gravity filter system employs three layers of filtration media: first, fine sand, followed by coarser sand and finally a layer of anthracite. The filter is cleaned by backwashing once a day and returns to its original form because of the differences in particle size. Other types of rapid filter may consist of a sand bed on top of a bed of gravel. Most modern systems are of the rapid gravity type. Slow filters, however, because of the time taken, offer the advantage of removing bacteria, algae and inorganic substances. Such a system is therefore suitable for the treatment of raw water supplies which have not undergone any other purification. Cleaning is carried out about every eight weeks by scraping off the top layer of sand and replacing it with fresh material.

After filtration the water may be passed through an *aerator* (either in the form of a fountain or water jets) to saturate it with oxygen before it is treated with chlorine. The chlorine dosage depends on the flow rate, and the sterilized water passes to the dechlorination unit: this is effected by the addition of sulphur dioxide. Some residual chlorine, however, may be desirable as a precaution against subsequent slight contamination.

Alternatively, sterilization may be effected by ozone, a practice favoured in Europe, but the cost of ozone production is high compared to chlorine and it is usually not economic for most countries. Ozone has the advantage that it needs no neutralization after sterilization.

The final process in many countries is fluoridation, a practice which has increased since the early experiments in Michigan, USA, in 1945. Fluoridation helps to prevent dental decay.

The purified water is transferred by powerful pumps to the mains supply.

In parts of the world with a shortage of fresh water supplies, modern desalination techniques have enabled sea and brackish waters to be transformed into fresh water.

Softening

Natural water supplies frequently exhibit a property called water hardness, owing to the presence of dissolved calcium and magnesium salts. It is these salts which react with soap to give insoluble precipitates, making it difficult to wash and resulting in the formation of a scum rather than a lather. Nowadays, however, the development of detergents has alleviated this, providing satisfactory washing conditions in hard water.

Water containing calcium bicarbonate and magnesium bicarbonate is said to be temporarily hard because the metal ions may be removed simply by boiling the water, resulting in the formation of insoluble calcium carbonate or magnesium carbonate. In hard water districts, such

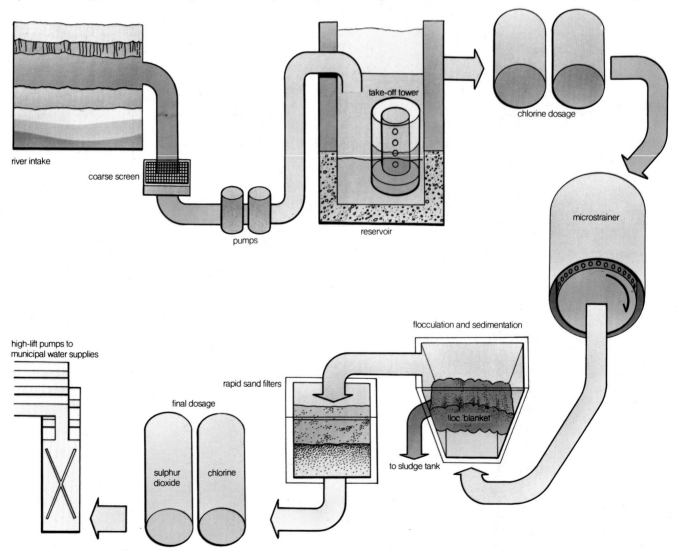

river intake

coarse screen

pumps

take-off tower

reservoir

chlorine dosage

microstrainer

flocculation and sedimentation

floc 'blanket'

to sludge tank

high-lift pumps to municipal water supplies

final dosage

rapid sand filters

sulphur dioxide

chlorine

deposits cause the 'fur' or scale in kettles. This has drawbacks in industrial applications, where the scaling of boilers and pipework results in inefficient heat exchange and may have other serious consequences.

Water containing calcium and magnesium in the form of sulphates and chlorides is said to be permanently hard because they cannot be removed as insoluble precipitates by boiling.

One method of softening water is by the addition of lime to precipitate the calcium ions as calcium carbonate and the magnesium hydroxide. Soda (sodium carbonate) is used to remove the sulphate ions and chloride ions by the formation of harmless insoluble salts.

An alternative technique is to pass the water through ion exchange resins, such as sulphonated polystyrene. Water flows through the resin and the calcium and magnesium ions in it are replaced with harmless sodium ions from the resin. When the resin becomes saturated with calcium and magnesium ions, it can be regenerated by flushing with a strong salt (sodium chloride) solution, the sodium ions on this occasion replacing the calcium and magnesium ions.

For water of exceptional purity, demineralized water, ion exchange resins are used in series: the first removes cations not by sodium ions but by exchanging them for innocuous hydrogen ions; the second removes anions, replacing them by hydroxyl ions. The resins are regenerated by flushing with an acid and a base respectively.

Above: a water take-off tower in Oxfordshire allows water to be taken off at any level where it is cleanest.

Left: a bulk lime tank. Lime is used to soften water by precipitating calcium ions.

Opposite page: what happens to water between the source and the user. Some cities, such as Manchester and New York, get their water from a distance through culverts from lakes; in this case it is already relatively clean. London gets it from the Thames and other rivers and the full treatment is necessary. It is distributed to users either by pumping it or by gravity feed. For gravity feed it is first pumped to elevated tanks; or in the case of artesian wells nearby hills may act as water towers.

Ships and docks

In 2500 BC the Egyptians were building fairly sophisticated sailing vessels; from then until the 19th century, ships were built mostly of wood and powered by sails.

At the end of the 16th century, the French inventor Denis Papin outlined plans for a boat with revolving paddles powered by a simple steam engine. Early steam engines, however, were far too big and heavy to be installed in boats. In the middle of the 18th century, the trickle of ideas, drawings and patents began to turn into a flood and finally in 1783 a paddle steamer called *Pyroscaphe* steamed against the current of the River Saone in France for fifteen minutes.

Three years later a steamboat built by John Fitch was tested on the River Delaware in America, and there were other early experiments. The *Charlotte Dundas* is known as the first successful steamboat because she towed two loaded vessels along the Forth and Clyde canal in March 1803. This steamboat was inspected by Robert Fulton, an American inventor and painter, whose *Clermont* began carrying farepaying passengers on the Hudson River in New York State in September of the same year, making the steamboat a commercial success.

Today nearly all merchant ships and warships are built of metal and powered by Diesel engines or steam turbines. The way in which the vessel is constructed depends upon the type of ship and the technique adopted at the shipyard, and this will be influenced by the available yard machinery and cranes.

For example, a bulk carrier will usually be constructed in the following way by most shipyards. Firstly the bottom shell and longitudinals will be laid on the building berth as a single unit after manufacture in the assembly shed, then the double-bottom unit will be lowered on to the bottom shell and welded into position. The wing-tank unit is lifted into position, aligned and welded up, and a pair of bulkheads are erected the correct distance apart over the hold length, with an allowance made for their inclination to suit the declivity (downslope) of the building berth necessary for launching. A side-shell panel can then be connected to the lower wing-tank unit and bulkheads to form the sides of the hold. Then the upper wing-tank is lowered into place and welded with the remaining deck panel finally completing the amidship structure. The ship is also built forward and aft of midships simultaneously. This technique, although not adopted by every shipyard, does allow an even spread of labour force. Working from midships gives a good reference structure for taking dimensions during the building.

When each heavy unit is lifted on to the berth, the bottom of the vessel is checked for alignment by an optical system. Any distortion which may occur would affect the strength of the hull as well as hydrodynamic efficiency.

Dry-cargo vessels

The basic orthodox design for a dry cargo vessel consists of a double bottom, several holds, a midship engine room and a forward and after peak-tank. Usually they have one or two decks and three main superstructures. These superstructures are a forecastle, bridge and a poop located at the bow, the middle and the stern of the ship respectively, and they extend to the sides of the vessel. The ship is sub-divided with steel divisions called bulkheads, which are watertight from the bottom of the vessel up to the main strength deck. Their main function is to restrict flooding if the hull is damaged, but they also support the deck and prevent the hull from distorting because of cargo or sea pressure.

The double bottom is a safety device in case the bottom shell is damaged; it also provides a space for storage of fuel oil, water ballast or fresh water. The double-bottom structure gives great strength to the bottom of the ship, which is essential for dry-docking operations. The forward and after peak-tanks are normally exclusively used for water ballast to give adequate draught when the vessel is unloaded and to adjust the trim if necessary.

The forecastle tween decks (short for 'between') are used for bosun's stores, the storage of wire ropes and rigging equipment and for paint and lamps. On the forecastle deck, each anchor cable passes from the windlass down through a spurling pipe into the chain locker, where the ends of the cables are connected to the fore-peak bulkhead by a cable clench.

At the after end there is a steering gear compartment where a hydraulic mechanism is used to move the rudder. The control for the steering gear is transmitted from the wheelhouse by a telemotor system. Directly below the steering gear compartment is the rudder trunk which houses the upper rudder stock that is used to turn the rudder. The poop and bridge are used for accommodation and for provision stores, some of which may be refrigerated.

As diesel machinery is thermally more efficient than other types it is often used in dry-cargo vessels. The propeller is driven directly from a slow speed, in-line engine. The propeller shaft passes to the after end through a shaft tunnel; this tunnel protects the shaft from the cargo in the holds and it provides access for maintenance of the shaft and bearings.

In addition to the main engine, the engine room contains auxiliary machinery such as diesel generators, oil purifiers, air compressors, ballast and bilge pumps, cooling water pumps and many other essential items of equipment. Just forward of the engine room are the settling tanks, fuel oil bunkers and a deep tank port and starboard, which may be used to carry liquid cargoes or a dry cargo such as grain or sugar. The accommodation is practically all amidships with the officers berthed on the bridge deck or boat deck. The wheelhouse, chart-

This Japanese-built supertanker was too large to be launched in the normal way, so it was built in a dry-dock which was then flooded.

room and radio room are usually together and the captain may have his dayroom, bedroom, toilet and office on the same deck. Galleys, pantries, lavatories and recreation rooms are carefully positioned to control the noise level and prevent annoyance to the off-duty crew.

The latest dry-cargo vessels have the engine room nearer to the stern. This shortens the shaft length and leaves a clear deck space forward of the bridge in which to work the cargo. Many vessels now have deck cranes for cargo handling instead of derricks operated by winches, and some vessels are fitted with special heavy lifting equipment.

Bulk carriers

These are single-deck, single-screw vessels which carry large quantities of bulk cargo such as grain, sugar, bauxite and iron ore. The engines are installed at the after end to leave the better spaces in the hull for cargo and the accommodation is all aft above the engine room, so that services and sanitation are concentrated in one region of the vessel. Upper and lower wing-tanks extend over the whole length of the cargo holds and they are used for water ballast when the ship is in the unloaded or light condition to give sufficient draught to immerse the propeller and give a better control over the empty vessel in heavy seas. The slope of the upper wing-tank is designed to restrict the movement of a grain cargo, which may otherwise cause the vessel to become unstable. The double-bottom tank is used for fuel oil or for water ballast, and these tanks can be used to make adjustments to the trim of the ship. Some bulk carriers have their own derricks or deck cranes, but many rely entirely on the dockside amenities for loading and discharging cargo. The hull construction of these vessels is a combination of two framing systems in order to obtain the best strength characteristics from each. The deck, wing-tanks and double bottom are longitudinally framed and the side shell is transversely framed.

Container ships

These vessels are a relatively new concept in cargo handling which reduces the time that the vessel stays in port. The containers also form a complete load for road vehicles without further handling. British built

The diagrams on the right are examples of structural features of common types of merchant ships.

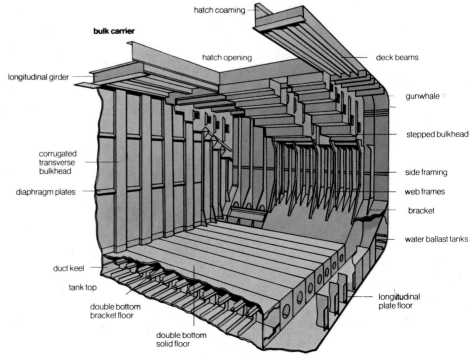

oil tanker midship section

upper deck
deck transverse
longitudinal bulkhead
longitudinal deck beams
gunwhale plate
deck girder
side transverse, forming ring girder system
oil tight hatch opening
face flat
web stiffening
upper horizontal girder
tie-beams
corrugated transverse bulkhead
centre tank
wing tank
longitudinal framing
lower horizontal girder
sideshell
bottom girder
bottom transverse
bottom framing

bulk carrier

hatch coaming
hatch opening
deck beams
longitudinal girder
gunwhale
stepped bulkhead
corrugated transverse bulkhead
side framing
diaphragm plates
web frames
bracket
water ballast tanks
duct keel
tank top
longitudinal plate floor
double bottom bracket floor
double bottom solid floor

Above: a large prefabricated section being lifted into place in a shipyard at Yokohama.

vessels are normally designed for 20 ft (6.1 m) long containers, but they can be modified for 40 ft (12.2 m) containers if necessary. The hold length is designed to suit the length and number of containers to be fitted into the hull, and to allow sufficient space for refrigeration coolers and coupling systems for these containers with perishable cargo. The accommodation and machinery on these vessels are usually located aft to leave a clear deck for cargo working and to allow the large crane an unrestricted region for operation. The shore container crane and its lifting spreader system will only lift standard containers and hatch lids with correctly designed corner fittings. All the holds have vertical guides to position the containers and give support, especially to the lowest container which could distort under the load transmitted down from those above. The containers are placed in a fore and aft attitude as the cargo experiences less ship motion in this direction, and when lifted ashore they are more readily received by road and rail transport. One advantage of container vessels is that they can carry containers on deck, but the number of tiers depends on the strength of the hatch lids and the necessity of having a clear view from the wheelhouse. The stability of a vessel with a deck cargo must always be checked, as the centre of mass of the vessel will be raised and it may cause the ship to roll or capsize. All deck containers are lashed to the hatches with steel rods or wires which have hooks and lashing screws to prevent the cargo being lost at sea.

LASH vessels

A lighter is a small barge which may be loaded with cargo. A LASH vessel is a mother ship which is capable of picking up loaded lighters at her stern and stowing them into large holds. (LASH stands for lighter aboard ship). The principle of the system is to collect together several loaded lighters into a rendezvous area with the LASH vessel ready for transportation over seas.

The LASH vessel has a single-strength deck, forward accommodation and a semi-aft engine room. The funnel uptakes are at the sides of the vessel to allow the massive gantry crane to pass down the deck on rails. Longitudinal bulkheads, steel divisions along the length of the vessel, and transverse bulkheads, steel divisions across the vessel, form holds within the ship to stow the lighters in cells. Vertical barge-guides are provided in the holds and the double bottom is equipped with sockets to receive the barge corner posts.

Walkways are provided with interconnecting ladders in the holds for the inspection and maintenance of the lighters. The gantry crane is supported at the stern by two large cantilevers; its lifting capacity is in excess of 500 tons and it is capable of transporting the barge along the deck to the hold. Each lighter is handled in about 15 minutes, and at present the LASH vessel will carry about 80 lighters, each with a cargo capacity of approximately 400 tons, a length of 61 ft 6 in (18.8 m) and a width of 31 ft 2 in (9.5 m). As well as being stowed in the holds, lighters can also be stowed in tiers of two on top of the large single-piece pontoon-type hatch covers, which have metal fittings for keeping the lighters secure during heavy weather at sea. The crane is equipped with a hydraulically operated latching

capable of preventing the spread of a flame in a 30-minute standard fire test, and the accommodation must have an automatic fire alarm and detection system.

A gyroscopically controlled set of stabilizers or fins are a common feature on most passenger vessels, to control the amount of roll and give a more comfortable crossing. For manoeuvring, these vessels are often fitted with bow thrusters, and they usually have twin-screw main propulsion.

The better cabins are located on the higher decks and the one, two or three-berth ordinary cabins on the lower decks. One of the most important areas in the accommodation is the foyer with reception desk, purser's office, main staircase and lifts. It should be centrally placed in order to receive the passengers so that their immediate needs can be dealt with as soon as they embark. The following public rooms are quite common on most vessels: restaurants, ballroom,

Laying the wooden deck in the construction of Cunard's liner Queen Elizabeth II.

device to grip the lighters and hatch covers, and a swell compensator which holds the lighter steady at the stern irrespective of the relative movements of the ship and the lighter in the sea.

The advantages of the system are that cargo-handling operations can be carried out in parts of the world where large ships cannot be berthed as the depth of water is insufficient. Mixed cargoes can be handled simultaneously, and the lighters can be towed to various places up river after unloading, thus providing a virtual door-to-door service.

Passenger vessels

In recent years the number of very large passenger liners has diminished in favour of the smaller vessel, capable of being converted for winter cruising. The vessel will normally comply with the regulations of all maritime countries, including those of the US coastguard, allowing it to change to cruising at any time.

Passenger vessels are more comprehensively sub-divided than other merchant ships so that, if several adjacent compartments are flooded, the ship will remain stable and stay afloat. If asymmetrical flooding occurs, the vessel has cross-flooding fittings to reduce the angle of heel.

Lifeboats are fitted port and starboard on the boat deck with sufficient capacity for the total number of passengers that the ship is certified to carry. Fire control is another important safety aspect, and the vessels are subdivided vertically into fire zones with steel bulkheads. In these zones the bulkheads must be

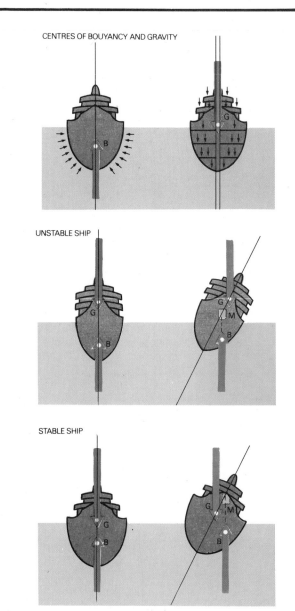

The relative positions of a ship's centre of gravity and buoyancy determine its stability. When the ship heels over, the line of action of buoyancy meets the centreline at the metacentre (M).

oil tanker

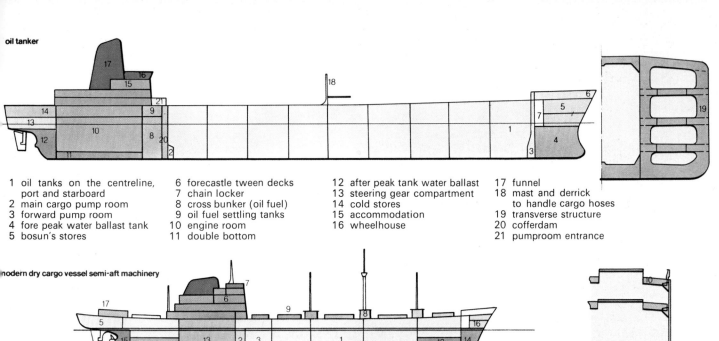

1 oil tanks on the centreline, port and starboard
2 main cargo pump room
3 forward pump room
4 fore peak water ballast tank
5 bosun's stores
6 forecastle tween decks
7 chain locker
8 cross bunker (oil fuel)
9 oil fuel settling tanks
10 engine room
11 double bottom
12 after peak tank water ballast
13 steering gear compartment
14 cold stores
15 accommodation
16 wheelhouse
17 funnel
18 mast and derrick to handle cargo hoses
19 transverse structure
20 cofferdam
21 pumproom entrance

modern dry cargo vessel semi-aft machinery

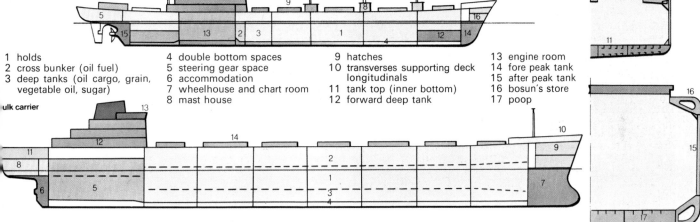

1 holds
2 cross bunker (oil fuel)
3 deep tanks (oil cargo, grain, vegetable oil, sugar)
4 double bottom spaces
5 steering gear space
6 accommodation
7 wheelhouse and chart room
8 mast house
9 hatches
10 transverses supporting deck longitudinals
11 tank top (inner bottom)
12 forward deep tank
13 engine room
14 fore peak tank
15 after peak tank
16 bosun's store
17 poop

bulk carrier

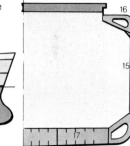

1 holds
2 upper wing tanks
3 lower wing tanks
4 double bottom tanks
5 engine room
6 after peak water ballast tank
7 fore peak water ballast tank
8 steering gear space
9 bosun's store
10 forecastle tween decks
11 cold store space (provisions)
12 accommodation
13 chart and wheelhouse
14 hatches
15 side frames
16 longitudinal framing in the upper wing tank
17 longitudinal framing in the double bottom

container vessel

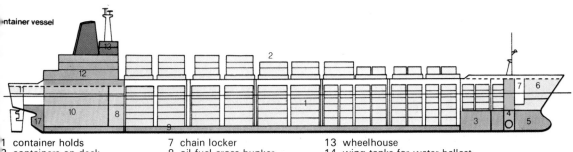
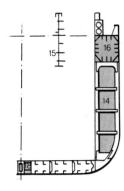

1 container holds
2 containers on deck
3 deep tank for trimming
4 bow thruster compartment
5 fore peak tank water ballast
6 bosun's stores
7 chain locker
8 oil fuel cross bunker
9 double bottom tanks
10 engine room
11 steering gear space
12 accommodation
13 wheelhouse
14 wing tanks for water ballast
15 longitudinal girder
16 pipe and cable passage
17 after peak tank water ballast
18 duct keel

LASH vessel (lighters aboard ship)

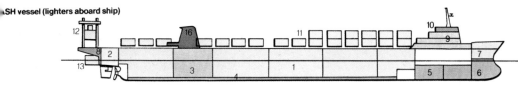
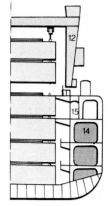

1 holds to contain lighters
2 steering gear space
3 engine room
4 double bottom tanks
5 deep tanks
6 fore peak tanks
7 bosun's stores
8 cantilever over stern for crane support
9 accommodation
10 wheelhouse
11 lighters on deck
12 crane to lift lighters
13 lighter at stern ready to be lifted aboard
14 wing tanks
15 passage way below deck
16 funnel port and starboard

163

cinema, discotheque, shops, cocktail bars, clubrooms, banks and hairdressers. For recreation there will be a swimming pool and a deck area for games, the young children will have a nursery and there are playrooms for older children. The officers are berthed near to the bridge and the remaining crew and stewards have accommodation on a lower deck.

Oil tankers

These vessels have a single main deck and a double bottom in the engine room only. Since tankers are divided into separate compartments, they are considered to be safe enough without having a double bottom along the full length of the ship. To reduce the risk of an explosion, the engines are fitted aft so that the shaft tunnel does not have to pass through any of the oil cargo tanks. At the extreme ends of the cargo tank range there is a cofferdam or bunker space to isolate the cargo from the other parts of the ship. Cofferdams are dry spaces across the vessel, preventing the possibility of any oil leaking directly into an adjacent compartment.

Pumps for discharging the cargo are fitted in a pump room in the bottom of the vessel. This pump room is often part of the cofferdam. The cargo pumps are usually driven by extended spindles from machinery in the main engine room. The oil is discharged from a tank by drawing it into the suctions at the end of a pipe leading from the main cargo pumps. It is then pumped vertically up from the pump room to the main deck where it passes along the deck pipelines until it reaches the deck crossover pipes. These crossovers are connected to the shore installations by hoses which are handled by the shore derrick. Oil tankers have small oiltight hatches with hinged lids giving access to the tanks by long steel ladders which reach to the bottom of the ship. The hatch coaming has a pipe leading vertically upwards to vent off vapour to the atmosphere if there should be a build-up of gas in the tank.

The oil tanks are subdivided by two longitudinal bulkheads into a centre-tank and wing-tanks, port and starboard. Sub-division of the oil tanks controls the movement of the cargo and prevents a large free surface across the ship which would allow the formation of waves in the tank generated by the motion of the ship. This wave will pass up and down the tank and could cause instability or structural damage unless controlled by wash bulkheads.

The engine power of a supertanker is very large, and a single propeller requires six blades so that the thrust is transmitted without overstressing the metal in any part of the blading. A bulbous bow is usually a standard part of the hull for a large tanker and it has the effect of modifying the flow of water at the bow, thereby reducing the power requirements from the ship's engines.

LPG vessels

Vessels of this type are designed to carry propane, butane, anhydrous ammonia and other liquefied gases in specially designed tanks. (LPG stands for liquefied petroleum gas). A typical gas tanker has a design similar to a bulk carrier but it has gas tanks built into the hull which rest on chocks and are keyed to prevent movement when the vessel is rolling or pitching. The liquid gas temperature in the tanks may be well below zero; this will cause severe thermal stressing when the liquid moves and therefore the tanks will alter in shape owing to temperature changes. A void space between the gas tanks and the hull is filled with an inert gas to prevent the oxygen in the air and any leak of gas from the tank producing an explosive mixture. An inert gas unit in the engine-room is used to produce sufficient gas for the void space and to keep it topped up in case of leakage. The tanks are made from a low-temperature carbon steel which must withstand impact at low temperatures and thus not be susceptible to brittle fracture. In some vessels the gas tanks are not refrigerated but are insulated with four inches (10.1 cm) of polyurethane foam. When the liquid gas vaporizes it collects in domes at the top of the tank where it is drawn off and passed through a reliquefaction plant and then returned to the tank via condensers in liquid form. The domes at the top of each tank project three feet (0.91 m) through the main deck.

Alternatively there may be no tank refrigeration or reliquefaction plant on the ship and the gas is free to 'boil-off'. The vapour is then transferred to the engine room and used as fuel for the main propulsion. This system makes the vessel less complex, cheaper on plant installation and bunkers but a percentage of the cargo is used over the voyage. Liquid-gas cargo is pumped from the bottom of the tanks using submerged pumps controlled from a room amidships. During pumping, a back pressure must always be maintained to prevent the gas boiling in the pump impeller and creating vapour in the riser when the tank is nearly emptied.

Air in the tanks is displaced before loading the liquid petroleum gas by using the inert gas system. Once the tanks are loaded they will always contain gas, so that the inerting procedure is not employed every time. Gas freeing of a cargo tank can be done by introducing inert-gas until the petroleum is diluted below the flammable limit and then blowing air into the tank and venting the gases at the top of the mast.

Docks

For many centuries sea traders relied solely on the shelter afforded by natural harbours, inlets and river estuaries in order to load or discharge, victual or repair their ships. While lying at anchor, their vessels were at the mercy, not only of wind and tide, but of bands of marauders to whom they were easy prey. The need for protection from such threats led to the establishment of basins or wet docks where sailing ships could be fitted out in safety and where their cargoes could be dealt with in relative security.

The word 'dock', which to this day is used fairly loosely to describe a variety of places where ships are berthed, was first used to describe 'an artificial basin filled with water and enclosed by gates' during the second half of the sixteenth century, a period of considerable expansion in maritime trade. One of the first recorded enclosed dock basins was the Howland Great Wet Dock which was built on the south bank of the River Thames in the seventeenth century. In the late eighteenth and nineteenth centuries the great dock-building period began in earnest, often closely associated with canal and railway-building ventures. This was also the period of the first iron steamships but the tremendous growth in ship sizes since those days has made many early docks obsolete.

The provision of gates at dock entrances is necessary because of the large tidal range which would otherwise cause the basins to have insufficient depth of water at low tide. In many countries the rise and fall of the tide is so insignificant that docks can be completely tidal. For example, in Melbourne, Australia, spring tides (those with the greatest range) rise less than 3 ft (1 m), in Rotterdam about 6.5 ft (2 m); and in Boston, USA, about 10 ft (3 m).

In Britain, however, with its large tidal ranges, most major dock systems are enclosed. The most notable exception to this is at Southampton where all dock berths are tidal, and where the effects of a 13 ft (4 m) tidal range are minimized by a phenomenon known as the 'double tide', which gives six hours of high water a day. An extreme example of a tidal range which makes enclosed docks imperative occurs in the Severn Estuary, where Bristol, for instance, experiences a maximum variation of almost 49 ft (15 m) between high and low water.

Lock entrances
The dimensions of the lock-pit inevitably govern the maximum size of vessel which can enter an enclosed dock. With the trend towards larger ships in recent years, the constraints of existing entrance locks have become a problem. The largest lock in Britain, at Tilbury, is 1000 ft (305 m) long and 110 ft (33.5 m) wide, with a depth of $45\frac{1}{2}$ ft (14 m), whereas the largest container ships, now operating between Europe and the Far East, are 950 ft (290 m) long overall, 106 ft (32 m) in beam, and have a maximum draught of $42\frac{1}{2}$ ft (12 m). New entrance locks are being built to cater for vessels even larger than this. At the West Dock at Bristol a lock measuring 1200 ft (366 m) long and 140 ft (43 m) wide is being constructed to take ships of 75,000 tons dead-weight, and developments at Le Havre in France include a new lock 1312 ft (400 m) in length and 219 ft (67 m) wide. This is claimed to be the world's largest lock and is capable of accommodating a tanker of 500,000 tons deadweight.

The operation of an entrance lock is basically simple. By using a system of culverts and sluices, water is allowed to pass from the dock into the lock with both inner and outer gates shut. The water level, and with it any ship in the lock, rises until it reaches dock level, when the inner gates open and the ship moves into dock. A departing vessel can then be penned in the lock and lowered by allowing water in the lock to escape through the outer sluices.

Dry docks and floating docks
At regular intervals all ships need to be inspected 'in the dry', and sometimes repaired. For this reason, most major ports are equipped with dry, or graving docks; slipways being used for smaller ships.

Dry docks, which usually take one ship at a time, are simply basins which are capable of being pumped dry to leave a ship supported by an arrangement of 'keel blocks', so that work can be carried out on the hull, propellers, or rudder. The procedure for drydocking a ship is a precise affair and may take several hours; with the dock flooded the gate is opened and the ship enters, then the gate is closed and pumping begins. Accurate positioning is vital as the ship settles on the blocks, prearranged to fit her hull, and to facilitate this, modern dry docks are usually fitted with guidance systems. In many international ports dry docks are being provided for the largest tankers afloat or planned: the port of Rotterdam already has a dry dock 1350 ft (412 m) long, which can accommodate 500,000-ton tankers, and the construction of a super dry dock for 700,000-tonners will be carried out at Kiel over the next two years.

The purpose of a floating dock is the same as that of a conventional dry dock, only the method of getting the ship out of the water differs. Ballast tanks are used to raise the submerged dock towards the surface and with it the ship to be repaired.

A large floating dock would have a lifting capacity of 20,000 tons, which would enable it to deal with ships of up to 70,000 tons deadweight. Such a dock, ordered recently for the Scandinavian port of Aalborg, will be 750 ft (228 m) long, 120 ft (37 m) wide between the inner walls and 148 ft (45 m) wide.

Container handling (below left) is replacing traditional methods of handling cargo by rope and sling. The picture looks down the boom of the container crane.

Surveying

Land surveying is the three dimensional measurement of natural and manmade features on the Earth's surface for representation in maps and plans. There are two basic types of surveying: *geodetic surveying* which takes into consideration the curvature of the Earth, and *plane surveying*, which assumes that the features to be surveyed lie on a horizontal plane. All large scale surveys are geodetic, but for areas with perimeter dimensions of less than 10 miles (16 km) plane surveying is usually accurate enough.

The first people to develop surveying techniques were the ancient Egyptians; a tomb at Thebes shows two men apparently surveying a field, and the pyramids could not have been built without fairly advanced surveying. The Romans used surveying extensively for fixing boundaries and marking out land for new cities, roads and aqueducts.

Surveying methods

The principles of surveying are based on the laws of simple geometry. If the positions of two points, say A and B, are accurately known and it is desired to fix the position of a feature such as a tree on the site being surveyed, any one of four techniques may be used: the distances from the tree to A and to B may be measured; the point at which the line between A and B passes closest to the tree may be located (often this is simply judged by eye) and its distance from A and from B measured; the distance from, say, A to the tree may be measured together with the angle between the line AB and the line from A to the tree; or, finally, the two angles made to the line AB by lines from A to the tree and from B to the tree may be measured. Fixing features by the first two techniques can be done simply by using a graduated steel measuring tape, and it is called chain surveying. The third technique requires the measurement of angles with a theodolite (see below) as well as distances, and is usually called theodolite traverse surveying. The final technique, called theodolite intersection surveying, involves only the measurement of angles.

Plane surveying assumes that all the various points fixed by measurement lie on a horizontal plane, and it is therefore necessary to adjust any dimensions taken up or down a slope to their plan projected length either by measuring the vertical distance between the two levels and applying Pythagoras' theorem or by measuring the angle of the slope with a theodolite.

A typical survey will be of a plot of land containing a house and a few trees. The surveyor will first choose a number of control points, say A, B, C and D, each positioned close to one corner of the site. Ideally, every control point will be visible from every other control point. The distances between the control points (including at least one diagonal distance between two opposite points) are then measured with a steel tape and plotted accurately to scale on a plan. The plotted quadrilateral ABCD forms a framework within which the other features of the site, such as the trees, the corners of the house and so on, may be located, usually by chain surveying. The

A self-reducing telemeter for measuring horizontally.

second of the two chain surveying techniques mentioned above, sometimes called perpendicular offset surveying, is generally chosen where there is much detail to plot on the plan. The surveyor will always try to build up a network of triangles with measured sides to cover the site, and for maximum accuracy he will keep the measured lengths as large as possible. Of course plotting the various features will be more difficult if the site is steeply sloping and the measured lengths have to be corrected to horizontal distances. If this has to be done, an optical level and staff will be used.

Levelling

An optical level consists of a telescope capable of rotation in the horizontal plane about a vertical axis. The telescope is mounted on an adjustable tripod and is fitted with a spirit level so that it can be accurately set up in the horizontal plane. Visible in the instrument's eyepiece is a horizontal line called the base reference; when the optical level is accurately set up, all points which appear on this line will be at the same level.

For establishing levels on a site being surveyed, the optical level will be used in conjunction with a staff, a rigid rule clearly graduated in metres and centimetres (or feet and tenths). The staff is first held vertically with its base on a bench mark (a point whose height above sea level is accurately known) or some other known fixed point. The optical level is set up some distance away (this distance will usually be as large as possible, but it will depend on the steepness of the slope) and focused on the staff. The reading on the staff where the base reference line crosses it is taken by observation through the level, and this gives the height of the instrument above the bench mark. The level is then rotated horizontally and sighted on to a second staff positioned on the point whose level is to be determined; a second staff reading is taken. By simple addition and subtraction, the height of the unknown point above the bench mark can be calculated, and this point can now be used as a fixed point for further levelling. In this way levels above or below a known level can be established wherever necessary on a site being surveyed.

Theodolite

If, in the survey discussed above, it is not possible to measure one of the diagonal distances (a building or other obstruction may prevent it), the internal angles between the sides of the quadrilateral ABCD will have to be measured, and this can be done either with a pair of compasses or, more accurately, with a theodolite.

The theodolite is an instrument for measuring angles in both the vertical and the horizontal planes. Like the optical level, it has a telescope and an accurate spirit level so that it can be set up horizontally, but in the case of the theodolite the telescope can be rotated about both horizontal and vertical axes which intersect at the instrument centre. Angular movements of the telescope about each axis are monitored on scales marked in fine angular increments. The accuracy will vary from instrument to instrument, but a good theodolite (though not of geodetic quality) will be able to measure angles to within about one second of arc, and this allows the surveyor to calculate accurate rectangular co-ordinates for points up to several miles apart.

Traverse surveying is used for all large-scale surveys; the theodolite is used to read all the angles between the various control points and all the distances are measured. It is possible to check the accuracy of such a survey by calculating around the traverse back to the starting point. For example, in the survey discussed above, after taking the necessary measurements, the position of B will be calculated from A, C will be calculated from B, D from C, and then A will be calculated from D: the closeness of the calculated position of A to its real position is then an indication of the accuracy of the survey as a whole.

Tacheometry

The infill detail for a large survey will be measured by whatever methods are most suitable, including *tacheometry* which is the optical measurement of distances. Most theodolites, in addition to the main vertical and horizontal sighting lines visible in the field of view, will have two short horizontal markings crossing the vertical sighting line, one above and one below the horizontal sighting line. These markings are accurately spaced so that when a staff is sighted, the difference between the upper staff reading and the lower staff reading against the markings will represent a fixed fraction (usually 0.01, one hundredth) of the distance between the instrument and the staff. Thus, if the lower marking falls on the one metre line of the staff and the upper marking falls on the two metre line, the difference between the readings will be one metre and so the distance between the instrument and the staff will be 100 metres.

This sort of tacheometry becomes more complicated when the line of sight is not perpendicular to the staff because there will be a foreshortening error in the staff

A geodometer uses a laser beam to measure distances up to 15 miles (25 km).

readings, and this has to be compensated for by trigonometrical corrections. The advantage of the technique is that the location of detail points and their reduced levels can be measured in one operation, but tacheometry by theodolite is not a very accurate method of measuring distances; errors of about 1% will usually be encountered.

The popularity of tacheometry has increased considerably in recent years because of the introduction of tacheometers. These instruments are theodolites with a mechanism for altering the optics to compensate for the angle of inclination of the sight line so that the horizontal distance between the instrument and its staff can be read directly. A common Swiss tacheometer has a split image arrangement in which the extent of the split varies with the inclination of the telescope. A horizontal, tripod mounted staff is sighted as the target, and an adjustment drum on the instrument is rotated until there is coincidence between markers on opposite halves of the split image. The horizontal distance between the instrument and the target may then be read to an accuracy of one part in 10,000 to a maximum distance of 220 metres (240 yards).

Accurate measurement of longer distances can be achieved with instruments which measure the time taken for electromagnetic radiation to travel to the target and back again. For distances up to 3 km (1.9 miles) light of known frequency, for example from a discharge tube or laser, is projected in a beam from the instrument to a reflecting target, and the time taken for the double journey is determined by comparing the phase of the projected light waves with that of the reflected waves. The frequency of the source is changed and the phase comparison repeated. A calculator incorporated in the instrument converts the phase differences into distance, this being displayed on dials or digitally. For distances up to 40 miles (64 km) a Tellurometer can be used, measuring the time taken for a radar signal to echo back from the target.

Triangulation

When a survey of a very large area is required, the surveyor will first establish a long base line and from that line locate a third point to complete the primary triangle. Typically, the length of the base line might be 20 miles (32 km), and its length would probably be measured by a Tellurometer. The primary triangle will form the basis for all subsequent measurements, and its three corners will constitute the main control points. This triangle will then be extended or broken down as necessary into smaller triangles.

167

Transportation of buildings

Houses have been moved for centuries. Henry VIII did it when he put together two smaller colleges at Cambridge to make Trinity College. Formerly, however, houses were nearly always moved by dismantling them and reconstructing them on the new site. Today it is often cheaper to jack up the house and move it without taking it apart.

Houses are moved because of road construction, new building projects, the clearance of areas that are to be flooded for new reservoirs, and other reasons. Moving houses is far more common in North America, where most houses are of wood frame construction, which is inherently strong and very light, and where roads are wider and straighter. In Britain, with comparatively narrow, winding roads, bridges, roundabouts and other obstructions, it is much more difficult to move a house any distance without considerable inconvenience. Even so, it is sometimes cheaper to move a house a short distance than to build a new one. If the house is of considerable historical or architectural interest, it may be moved to save it from demolition.

Each house-moving job must be studied carefully and the method of operation chosen depending on the structure of the house, the route over which the move will take place, and the nature of the new site. Thus the engineer must start at both ends of the problem. The construction of the house must be studied to determine how the different parts of the building are supported, and to calculate the weight supported by each part of the foundation. Then the route and the new site must be studied to determine whether a gradient exists, and if so how much; whether the ground is soft or firm and what obstructions are in the way; and whether the house can be moved in a straight line or must be turned through an angle.

The moving method will be either to install tracks on the route or to use a large wheeled or tracked transporter. Large buildings are rarely moved long distances and tracks will always be used; for smaller buildings the transporter may be a heavy lorry [truck] or purpose-built *bogies* made up of large numbers of closely spaced wheels.

Once the transport system has been decided attention can be turned to the building and the design of the chassis. A method of constructing beams of reinforced concrete beneath existing walls was developed in England in the 1940s; this method is best for brick or masonry buildings. If tracks are to be used to move the building, brackets are bolted directly to the beams; specially constructed hydraulic jacks are fixed to the brackets. The building is then jacked up between 18 inches and three feet (about 0.5 to 1 m) to allow for tracks to be run underneath the foundation at ground level. This is accomplished by thrusting off timbers or concrete pads placed beneath the beams during the construction. The jacks are extended to lift the building and temporary packings are placed under the beams on each side of the jacking positions; the jacks are retracted, packings placed beneath them and the process repeated until the house is high enough. This must be carefully planned so that support under the building is always equal at each point. When the house is high enough, the tracks are laid and wheels to run in them are fitted to the jacks.

If the building is to be moved on groups of wheels without tracks, the concrete beams can be smaller and lighter and supported themselves at more closely spaced intervals by a second frame of steel beams. This may be a type of construction similar to a temporary bridge erected around the building, which can be quickly assembled and quickly removed when the move is completed. In the case of timber framed buildings, the whole building can

be supported by a steel chassis, dispensing with the concrete beams. This may take the form of steel trusses carrying the load to Bailey-type bridging, or of a network of main steel beams resting on the wheels with smaller beams carrying the wall loads to the main beams. Smaller lighter buildings can be supported entirely by heavy timber. The advantage of timber is that it can be easily cut and fitted to the job and is lighter to handle.

The condition as well as the construction of the building must be taken into account. If the building is of brick or masonry and in good condition, the support must be absolutely rigid to avoid cracking. If it is an old timber building in poor condition which is to be restored after moving, a considerable amount of flexibility is allowed and the supports can be much lighter. If the building is to be moved on tracks, it is very expensive and time consuming to pack the ground to an absolute level to prevent movement of the track as the weight passes over; a thick layer of well-packed ground is needed underneath. A method which can be used to compensate for some local movement of the track is to support the building on jacks over each wheel, and join the jacks together in three groups, providing approximately equal pressure so that they act as a single three point support.

In Suffolk in 1972 an Elizabethan manor house 90 feet long and 40 feet wide (about 27 × 12 m) was moved ⅝ mile (1 km). Planning permission had been given for a housing and commercial estate provided the house, scheduled for preservation, was moved bodily to a new site over the brow of a hill so that the new housing estate would not be in view. At one point the house was hauled up a hill which inclined the building ten feet (about 3 m) out of level. A steel frame was installed inside the house to strengthen it and the load was carried to a Bailey bridging framework around the outside. The house was jacked up nine feet (2.7 m) to allow for a chassis and transporter system made up of a line of fourteen wheels at the rear, grouped to balance the load, and separately steerable groups of six wheels each on the front corners, with smaller steering bogies in front. The structure was winched over unprepared ground using pulley blocks anchored into the ground, and lowered into place on the other side of the hill, 60 feet (about 18 m) higher than the original site.

A brick house was moved in Staffordshire recently, using reinforced concrete chassis beams with brackets, jacks and wheels running in tracks. The jacks were linked in three groups and the building was moved three hundred yards (274 m) including a right-angled turn, without any cracks appearing in the brickwork.

Left: a 380-year-old house moved on a steel framework.
Below: this brick house was moved on tracks. Even the creeping plants were moved with it.
Bottom: a wood frame house is like a box and fairly easy to move; this one, in New Jersey, was sawn in half.

Tunnels

Tunnels have been known for almost as long as man has been digging holes in the ground. One of the oldest tunnels known for the sole purpose of transportation was constructed some 4000 years ago to pass beneath the bed of the river Euphrates. The scale of the work indicates that the engineers of the time must have acquired considerable expertise in tunnels built even earlier, but nothing is known of these earlier constructions. Various other tunnels are recorded as being constructed at an early date in the Middle East. By 600 BC tunnels had been driven from both ends for distances exceeding 600 yards (550 m) showing a skill in survey that would be acceptable today.

Tunnels can either be bored, or built by the 'cut and cover' method, that is digging a trench and then filling in part of the excavation to form a roof over the lower part of the trench. Originally, tunnels were driven through ground that was self supporting, since the techniques for supporting weak ground such as soft clays and gravels below the water table level were undeveloped.

The canal era in Britain in the 18th and early 19th centuries involved the construction of many major tunnels with temporary timber supports and permanent brickwork linings. With the coming of the railways, larger sections of tunnel had to be constructed and the temporary supports in soft ground became more and more complex, in some cases even hampering the erection of the permanent lining. This situation called for new methods of support, which have become standard features of soft ground tunnelling used worldwide.

The most important development was the invention by Marc Isambard Brunel of the tunnel shield for the construction of his tunnel under the river Thames in England between 1827 and 1842. This shield was rectangular in section and was advanced by screw jacks thrusting off the front of completed lining, which was made of brick. Modern tunnel shields owe their inception to G. H. Greathead, who designed a circular shield for another tunnel beneath the Thames in 1869, this time driven through non-waterbearing clay and lined with cast iron, another important innovation.

The shield is a movable frame, generally cylindrical, which supports the face of the tunnel and the ground immediately behind the face, affording protection to the tunnel miners who carry out both the excavation of the face and the erection of the lining. It consists of an outer envelope, or skin, of steel, somewhat larger than the external diameter of the tunnel lining, which is stiffened internally by diaphragms or, in the case of larger shields, with a heavy framework of structural steel. Mounted integrally at the leading edge of the skin is a thick cutting edge of steel, which is pushed into the face and trims the excavation to the shape of the tunnel. A series of hydraulic jacks are mounted on the rear of the shield inside the skin, and these are used for pushing the shield forward as each successive length of tunnel is excavated. Towed behind the shield is a series of sledges which carry all the paraphernalia of modern tunnelling, principally dirt handling and grouting equipment.

Methods of tunnelling

Activities in tunnelling have tended to divide into two distinct fields: hard ground tunnelling and soft ground tunnelling. In each case, excavation of the tunnel face is carried out both by men and by machines, although the techniques required for each field vary widely.

When tunnelling through hard ground without a tunnelling machine, the rock face is drilled by pneumatically or hydraulically operated drills so that explosives can be inserted into the drill holes and detonated with time delays to give the desired fragmentation to the rock debris and profile to the tunnel. Where space allows, the drills are mounted on a drill carriage, or *jumbo*, thus allowing the drill operator to control a number of drills at the same time. The pattern of drilling is carefully controlled to provide the maximum length of 'pull' (depth of excavation) for each detonation. Generally, two methods of drilling are adopted, the wedge cut and the burn cut. In the wedge cut, the holes are drilled in a converging pattern towards the centre of the face so that when the charges are set off the rock is blown back towards the centre of the tunnel. The burn cut, where a large central hole is drilled in the tunnel face and the remainder of the

A contemporary illustration of Marc Isambard Brunel's tunnel shield, used from 1827 to 1842 to build the first tunnel under the Thames.

The CAP (comprehensive all purpose) tunnelling machine is suitable for hard and soft work. Hydraulic power does most of the work; note the bank of valves at the left.

holes are drilled parallel to the longitudinal axis of the tunnel, requires less skill than a wedge cut and greater lengths of pull are usually obtainable, but it consumes more explosive.

In both cases, the pattern of rock fragmentation is similar. The explosions are timed so that the blast commences in the centre of the tunnel face, creating a space into which rock is blown by succeeding charges. As the blast progresses, concentric rings of rock are blown inwards and the central area, although never larger than the volume of broken rock being forced into it, provides sufficient space for the debris to accumulate.

The techniques of tunnelling in soft ground evolved from mining methods and consisted of placing a series of boards, or runners, closely around the perimeter of the tunnel face. These were driven forward into the ground as a form of horizontal sheet piling suitably strutted and braced. The face of the excavation itself had to be kept closely boarded with horizontal boards which were removed one at a time, and a small area of ground was then excavated so that the board could be replaced a short distance ahead of the face and secured. This process was repeated until sufficient space had been created to insert a fresh length of runners and strutting. Progress was obviously very slow since considerable care had to be exercised at every step to prevent movement of the ground in any direction. In spite of this, however, accidents did occur and roof falls were frequent.

Mechanization

The earliest attempts to mechanize tunnelling were made about the end of the 19th century. It was at this time that Colonel Beaumont invented a machine for tunnelling through chalk for his attempt on the first tunnel between England and France. The tunnel was abandoned because of political and military considerations rather than technical problems after about a mile (1.6 km) had been dug beneath the sea. It is interesting to note that the line of the most recent Channel tunnel project intersected the Beaumont tunnel, thus showing that his choice of route so many years ago was technically sound.

Tunnel excavating machines are of two kinds: full face and partial face. The full face machine must necessarily be circular, since the entire front face of the shield rotates. Depending on the hardness of the ground, the arrangements for excavation vary. In soft ground, plates pressing against the tunnel face may be used; these are adjusted to allow only a certain amount of ground to be excavated at any one time. For harder ground and rock, individual teeth or cutting discs or rollers may be used. The hardness of the ground is not the only criterion that dictates the choice of teeth; abrasiveness also has to be taken into account.

More recently, full face mechanization has been applied to the process of tunnelling through waterbearing sands and gravels with the introduction of the bentonite tunnelling machine. This technique makes use of the clay

mineral bentonite in a sealed pressure chamber in front of the shield to provide a stabilizing effect on the unstable ground at the tunnel face, while at the same time allowing mechanical excavation to take place at a rate much faster than was previously possible by hand. The excavated material mixed with the bentonite slurry is extracted from the pressure chamber and then taken to the surface by hydraulic pumping. On the surface, separation of the bentonite from the sand and gravel takes place so that the bentonite can be recirculated, and the sand and gravel is taken away in trucks.

Partial face mechanization takes the form of rail mounted machines with a cutting head attached to a hydraulically controlled boom which can range over the entire area of the tunnel face. In some cases the cutting boom is mounted within a tunnel shield, the cutting profile being controlled by a guide ring in the shield.

There are many ways of removing excavated material from the tunnel face. In small tunnels in both hard and soft ground, railway wagons on narrow gauge track are frequently used and they are loaded by hand, conveyer, or mechanical loaders powered by compressed air. As the size of the tunnel increases, and where the floor is reasonably flat, conventional diesel engined dump trucks are used, and the loading of these trucks will usually be done by front end mechanical shovels which combine high manoeuvrability with a large output.

Tunnel linings

Of the various considerations governing the choice of lining for a tunnel, one of the most important is the need to withstand any load imposed on it during and after erection. If a shield is to be used, the lining must be capable of withstanding the thrust of the shield and also any other stresses induced by handling. It should have a

high strength to weight ratio so that its thickness can be minimized to save unnecessary excavation.

Originally, linings were made of wood, and then brick, but the most important innovation in soft ground tunnelling was the introduction of segmental cast iron at the end of the last century. Today, however, cast iron is expensive and the alternative of pre-cast concrete is frequently preferred. In hard ground tunnelling, concrete is most often used when a lining is required. Factors which must be taken into account when designing tunnel linings include water and ground loadings, rock loosening loads, the construction of subsequent tunnels and any special characteristics of the ground. The time dependent response of the ground to the actions of tunnelling must also be considered. The initial process of excavating a hole in the ground is followed by the erection of a structural lining. The time interval between these two operations greatly affects the loads which the lining can carry, and it is often important to know how long a tunnel can safely remain unlined.

Specialized techniques

The presence of water in the ground has a tremendous effect on the construction of tunnels, and techniques

employing compressed air, ground injection and freezing are available, particularly when tunnelling through soft ground.

Although the principle of compressed air as a means of excluding water from diving bells had been known for centuries, it was not until the last century that James Greathead first used compressed air in tunnelling operations. Previously the development of compressed air methods had been delayed by lack of sufficient technology. The air lock, permitting the passage from one pressure to another, was not invented until 1830.

The main reason for using compressed air is to prevent water from entering the tunnel, but it also provides an important function in increasing the load on the tunnel face, in some cases allowing a reduction in the quantity of timbering. There are risks in using compressed air since nitrogen in the air is forced into the blood under pressure and if decompression is uncontrolled, the nitrogen forms bubbles in its attempt to leave the bloodstream, causing what is known as 'the bends'. Each country has its own standards, but the use of compressed air is now governed by strict regulations to ensure that both compression and decompression are performed in a safe and efficient

173

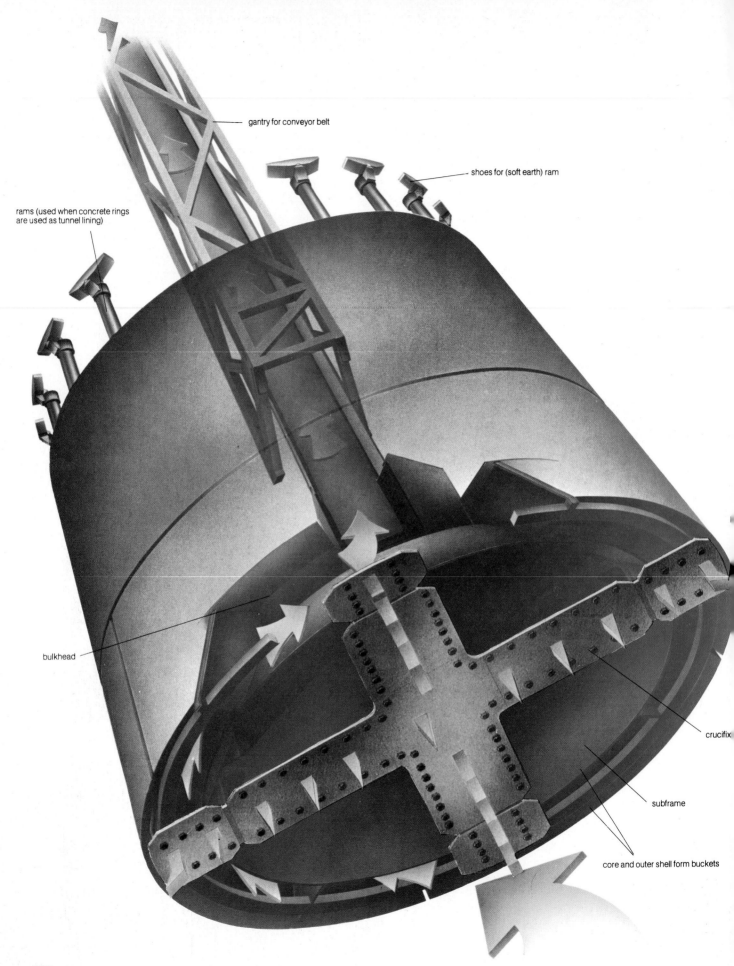

gantry for conveyor belt

shoes for (soft earth) ram

rams (used when concrete rings
are used as tunnel lining)

bulkhead

crucifix

subframe

core and outer shell form buckets

manner, and the incidence of all kinds of compressed air sickness has been sharply reduced.

In addition to compressed air, ground treatment can also be used to reduce the flow of water into a tunnel, and it is now possible to inject all but the most finely grained ground materials with cement or chemical grouts which form a matrix to fill the voids between the soil particles and bind them together. The injected materials range from cement and clay for filling the larger voids to polymers and low viscosity grouts for fine sands, the cost depending to a large extent on the material injected. The work can be carried out either from the surface or from within the tunnel itself.

Refrigeration techniques have also been employed since the first freezing plant was set up for shaft sinking in 1883. The temperature of the soil is lowered by drilling or driving tubes into potentially unstable soil and then circulating a low temperature liquid such as cooled brine or liquid nitrogen. The water in the soil gradually freezes around the probes and ultimately ice 'cylinders' build up to form a continuous wall of frozen ground which forms an effective barrier against soil and water entering the tunnel.

Opposite page: a tunnelling machine. The rotary cutting head is driven by four electric motors, while 16 hydraulic rams push the head forward. The thrust is co-ordinated with the head so that if it stalls the forward thrust is reduced.
Below left: the Paris Metro in 1903.
Bottom left: the cut-and-cover method for a road under Dubai Creek in the Persian Gulf.
Below: cut-and-cover in Tashkent for the first metro in Central Asia.
Bottom: a station on the Moscow Metro.

The Vault

Since the principal reason for building is to enclose space, the chief problem for the architect or builder has always been the construction of the roof. Wherever wood was abundant this was always the easiest material to use; vaults of stone or brick were first evolved in countries where wood was scarce. Later they were felt to be desirable in well-wooded countries too, because of their greater strength, finer appearance and better protection against fire. But vaults in European countries are nearly always protected from above by wooden roofs covered usually with stone slates, tiles, lead or copper.

Barrel and groin vaults

The simplest kind of vault is a tunnel or half-barrel, round-arched and borne on continuous lengths of wall. Vaults of this kind were constructed by the Assyrians and by the Romans—on an enormous scale in Rome itself, thanks to the presence locally of materials (a sandy earth known as *pozzolana*, and lime) from which a very strong concrete could be produced. Roman vaults were usually coffered: coffers are sunk panels, square or polygonal and often in diminishing perspective, introduced not only for

ornamental purposes but also to reduce weight.

When two tunnel vaults of equal radius were constructed to intersect at right angles, the outcome was a *groined* vault, which was also used by the Romans; this is the characteristic vault of Romanesque architecture. Its practical advantages were two: the concentration of the thrust at the four corners of each vault-bay rendered possible opening into side aisles and the insertion of windows above these openings to improve the lighting. But so long as all the arches remained semicircular, proportions were rigid: each bay could only be an exact square.

Gothic vaults

The greater contribution made by the introduction of the pointed arch was flexibility. If the apex of the tunnel is lifted to a point, it can be intersected at right angles by a narrower tunnel climbing to a sharper point: the resultant vault-bay will be rectangular. It was soon realized that a bay of any reasonable shape could be vaulted, simply by varying the sharpness of the angles of the various arches. This basically simple discovery was exploited by the

Gothic architects with infinite resourcefulness.

The other innovation at the vault mainly associated with Gothic architecture is the appearance of ribs, which seem to strengthen the groins. It used to be believed that the whole weight of a Gothic vault was carried by the skeleton of ribs; that this is an error was proved in World War II, when at various places ribs were destroyed without any subsequent collapse of the vault. The function of the ribs, in fact, is not structural but aesthetic.

As the Gothic age advanced, however, vaults were constructed with ever increasing skill: this implied not only finer masoncraft but, above all, deeper understanding of how to counteract the thrusts which a vault exerts. These are twofold: downwards, through the weight of the material itself, and outwards, in an oblique direction, through the pressure of the wedge-shaped blocks with which every stone vault is built. The downward thrust was quite a simple matter: this was carried by the piers, which, as masoncraft improved, were able to become more and more slender. The outward thrust posed a much more difficult problem. It was solved by the development of buttresses, and, as the churches grew loftier, of flying buttresses, even, at the boldest French cathedrals, in several tiers. The concentration of the whole thrust of each vault-bay at the four corners meant that the intervening sections of wall, so essential to the tunnel vault, now served no structural purpose. Hence the gradual removal of the walls and substitution of windows, until in the late-Gothic period almost every church seems to be 'more glass than wall'.

The first vaults, whether groined or ribbed, were *quadripartite*: that is to say, they comprised four compartments of equal size. In early Gothic vaults an additional transverse rib (the rib set at right angles to the axis of the portion of the building in question) was sometimes introduced, thus dividing the vault bay into six compartments (of unequal size) instead of four: this is known as a *sexpartite* vault, and is mainly to be seen in France, but occurs also at Canterbury Cathedral in England.

Subsequent developments were all ornamental rather than structural. First came the *tierceron* vault. Tiercerons are pairs of ribs which have the same point of springing as the principal ribs but which meet at the ridge obliquely instead of being carried across from one side of the vault to the other in a continuous line. Such a vault is seen to perfection at Exeter Cathedral.

Lierne vaults did not appear before the fourteenth century. Liernes (from the French *lier*, to tie) are short ribs which neither spring from the capitals nor, in many cases, rise to the central ridge; but by crossing and recrossing the more functional ribs it was possible, with liernes, to make patterns—stars, for example—of great decorative beauty.

Still more original and ingenious was the *fan* vault, a purely English invention, and never better used than in one of its earliest manifestations, the cloisters at Gloucester. Its essential feature is the inverted half-conoid: that is, a half-cone with concave sides. Each pair of half-conoids just touch at the centres of their curves. By comparison with the tierceron vault, from which it

Far left: the earliest vaults, in tombs at Ur and in the Pyramids, were vaulted by corbelling or overlaying the stonework so that it jutted out. This method was copied here by the Etruscans for their tombs.

Centre left: the nave of the abbey church at Veselay, France, c.1120, an early example of straight groins over rectangular instead of square bays. In spite of this elongation, the groins are in effect elliptical arches. Later they began to be thought of as four arcs meeting at a common centre.

Left: the next stage was to emphasize the groins by making them into ribs, as seen here in the cellarium of Fountains Abbey, North Yorkshire, founded in 1132.

obviously evolved, the fan vault was a step towards standardization; the ribs are not applied separately but carved out of the surface of the solid block of stone; they are all equidistant from one another, and usually their curvature is identical. In fan vaulting it is the central spandrel which is the problem. Here the curvature is almost flat, and with wide spans, in order to avoid its becoming dangerously large, and therefore unsafe, the only thing to do was to slice off the sides of each fan, instead of carrying them round the full 180 degrees. Aesthetically, as can be seen at King's College Chapel, Cambridge, this is a pity, wonderful as the vault is.

The final feat of late Gothic virtuosity was the *pendant* vault: there are both pendant liernes (Oxford Cathedral) and pendant fans (Westminster Abbey: Henry VII's chapel). These were achieved by elongating two of the *voussoirs* (wedge-shaped blocks of stone) of which each transverse arch is composed: the pressure of the others, properly counteracted by piers and buttresses, holds them in position. Each pendant is carved out of a single block of stone, which could be 10 ft (3.5 m) long. Not only was the pendant vault, in a girderless age, a feat of extraordinary architectural ingenuity; it was also a triumph of artistry.

Construction

Vaults were built on strong temporary framework known as *centering* which consisted of timber framing in the

The cloisters of Gloucester Cathedral, begun after 1351, offer an exquisite example of a purely English phenomenon, the fan vault. The pattern of ribs of equal length and curvature form inverted half conoids.

form of the arches and was the carpenter's responsibility. Once the centering had been built and while still on the ground, the stonemason cut the stones to shape. They were then numbered so the correct few could be winched up as needed to the masons working high on the scaffolding above.

The centering was raised above and the wedge-shaped arch stones, or *voussoirs*, were assembled on it with thick mortar joints. Finally at the apex, the large keystone was dropped into place, weighing the ribs against any tendency to rise, but it was the 'shell' characteristics of the web itself that gave the vault its strength. The compartments defined by these ribs were then filled in with the *web*. First, it was necessary to secure planks or lagging between the formwork to support the web. The mason stood on a platform below the vault and, working from the side, laid on one course of stones above the other; his masonry load being balanced by that of another mason working simultaneously on the opposite side of the vault. Because of lack of space, the last few courses were laid by one man. Sometimes the web stones rested on the back of the ribs, but another technique was to cut rebates into the rib so that the stones rested on these, separated by the *rib stem* which projected upwards between them.

Web thickness appears to have varied from a very thin 'shell' of about 3 inches (7.6 cm) to 10 inches (25 cm) or more. At one stage vaults were overlaid with concrete to strengthen them, but this was not necessary as by their very nature they were strong enough.

The underside of some vaults was plastered and might also be covered with paintings as at Saint Savin-sur-Gartempe near Poitiers, France.

It is not certain how mediaeval vaults were built, but collapsible framework such as shown in (a) is a likely method. In (c), six types of vaults are described, the earliest first. In (d), the groin vault is like two barrel vaults intersecting, which is how it originated. The framework for a barrel vault could be moved along by removing a spacer block, but had to be completely dismantled and re-erected for the groin vault. In later vaults (e) the solid formwork built on from above was abandoned and the stones hauled up from below by a treadmill-operated crane or by a windlass. The masons built towards each other until they met at the top.

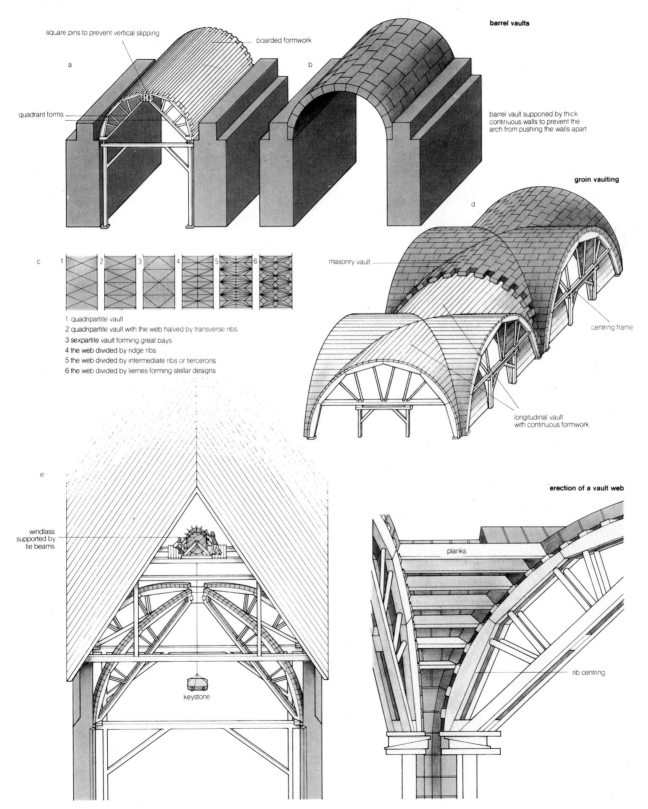

square pins to prevent vertical slipping

boarded formwork

a

quadrant forms

barrel vaults

b

barrel vault supposed by thick continuous walls to prevent the arch from pushing the walls apart

c 1 2 3 4 5 6

1 quadripartite vault
2 quadripartite vault with the web halved by transverse ribs
3 sexpartite vault forming great bays
4 the web divided by ridge ribs
5 the web divided by intermediate ribs or tiercerons
6 the web divided by liernes forming stellar designs

groin vaulting

d

masonry vault

centring frame

longitudinal vault with continuous formwork

e

windlass supported by tie beams

keystone

erection of a vault web

planks

rib centring

Wells

A reliable water supply is perhaps the most important consideration for any human habitation. In many places a river or stream can produce an adequate supply, while others may rely on the natural flow of water from the ground at a spring; but the artificial tapping of ground water by means of a vertical shaft is often the only means of providing a completely reliable source of water. Up until 1930, for example, the city of Teheran (in Iran) was completely dependent upon twelve wells for its water supply; and in many desert regions human settlements are only possible because of this subterranean water source.

The term 'well' can also be used for a vertical shaft which produces other liquids; oil, sulphur that has been melted by hot water forced down the well into a sulphur deposit, or salt solution from water forced into a salt deposit.

Water wells depend on the fact that many rocks are permeable to the flow of water. Rain falling on to rocks such as limestone, sandstone, basalt, gravel or sand sinks through until it reaches the level at which the rocks are saturated with water. This level, the *water table*, varies in depth from over 100 ft (30 m) in arid regions, to ground level in some localities. At these places water appears at the surface, either as a flowing *spring* or a non-flowing *seep*. A vertical shaft sunk to below the depth of the

water table will fill with water until its surface is at the level of the water table in that locality, because of seepage from the surrounding rocks. Traditional methods of 'drawing' water from a well include the bucket and windlass for a moderately wide (3 to 6 ft; 1 to 2 m) brick-lined well, and the 'village pump' which abstracts water through a pipe a few inches in diameter. Many wells, however, have a water level which is only a few feet below ground level and can be reached by a short descending flight of steps.

The wells discussed so far are termed *shallow wells*, as opposed to the deep *artesian* wells mentioned below. All shallow wells are susceptible to contamination, particularly by sewage, as in the case of the pollution of a well in Broad Street which caused the London cholera epidemic of 1854. The risk is decreased by lining the well with impervious materials, and making it as deep as possible so that water from a great depth seeps in. A greater depth also means that the well is less likely to run dry during a drought, when the level of the water table falls.

Artesian wells

In the London region, the strata of chalk form a 'saucer', which rises at the edge in the Chiltern Hills and the North

Primitive well machinery at Djerba, Tunisia.

Downs, and is covered in the centre by the impervious London clay. Before the large-scale sinking of wells (up to 1000 ft (305 m) deep) in the London region, rainfall on the hills produced a water table there which was higher than the ground level in London, so a well sunk through the clay to the chalk would actually flow because of the pressure of water in the rock. Such *artesian wells* still contribute to the water supply of London, although the ground water has now been so depleted that the water table lies below the London ground level and the water must be pumped out: it is now, strictly speaking, *sub-artesian*.

The name derives from the French province of Artois, where artesian wells were in use as early as the 12th century; another artesian basin lies under Paris, but probably the largest is Queensland. The Great Artesian Basin, as it is known in Australia, covers an area of over 600,000 square miles (1,500,000 sq km), and some of the wells are 4600 ft (1400 m) deep. At this depth the temperature is near the boiling point of water, and it must be allowed to cool before it is given to cattle. This supply has too high a mineral content for use in irrigation, which is the principal use of well water on a worldwide scale.

Above: an artesian well in an orange grove in southern California, where such wells are a main source of water.

Below: if the water table lies below the well, it acts as a water tower, providing gravity feed.

depression of water table
due to high pumping rate

well refills more quickly if provided
with horizontal adits below water table

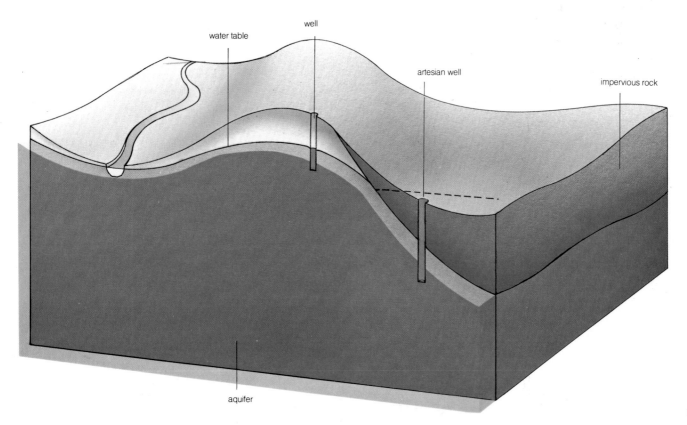

water table

well

artesian well

impervious rock

aquifer

Index

Entries for chapter
headings are in
capital letters.

aeration *see* SEWAGE
aftercooler 55
air hammer 52
alluvial mining 121
Ambasz, Emilio 98
American Motors 14–15,
 20–21
arc welding 17, 30, 45
ARCH, the 9–11 *see also*
 BRIDGES
architecture *see* ARCH,
 BUILDING,
 CATHEDRAL,
 DESIGN, VAULT,
 etc. *see also* Byzantine,
 English, Gothic, Greek,
 Renaissance, Roman,
 Romanesque, etc.
Argand, Aime (burner) 116
artesian wells 180–181
Aspdin, Joseph 44, 78, 82
asphalt 84, 85
ASSEMBLY LINES
 12–21
Assyrians 10
auger mining 124
Austin cars 13

Babylonians 10, 113
backhoe *see* excavating
 machines
Bauhaus 94, 97
Beaumont, Colonel 171
beer 123
Bessemer process 32
bitumen 85–86, 143, 145
BOATBUILDING
 22–27
boom *see* crane
bricks 39, 46, 74–77
BRIDGES 28–38
bronze, bronze age 39, 59,
 133–137
Brunel, Isambard
 Kingdom 29, 44
Brunel, Marc Isambard
 170
BUILDING, history
 39–43 *see also* DESIGN
BUILDING, modern
 44–48
buildings (moving them)
 168–169
bulldozer 67–69

bund *see* dam
buttresses 9, 177 *see also*
 CATHEDRAL
Byzantine, Byzantium
 39–40, 41

Cadillac 20–21
caissons 30
canals 170
canoe 22
car assembly *see* assembly
 line
carvel 23
CATHEDRAL
 construction 49–54 *see
 also* BUILDING,
 history
cement 39, 78–85
centrifugal force 59
Charlemagne 40
channel tunnel 171
Charlotte Dundas 158
Chevrolet 17
Clean Air Act (1956) 123
Clermont 158
clincher, clinker 23
coffer-dams 29
Colosseum 10
Colt, Samuel 12
composting 150
compressors (air),
 compressed air 13, 20,
 30, 55–56, 61
computers 16, 107–108
Concorde (aircraft) 12
concrete 11, 26, 80–85, 112,
 143–145 *see also*
 BRIDGES and
 ROADBUILDING
CONSTRUCTION
 MACHINERY 55–73
CONSTRUCTION
 MATERIALS 74–86
conveyor 12, 13, 14, 16,
 18–19
corbelled ('false') arch 11,
 39
cranes 37, 47, 52–53,
 63–66
cruck (house) 42
Cunard 162
curragh 22

Darby, Arbaham 44
dams 105–108, 113–114
DEMOLITION 87–89
derrick 64
DESIGN 47, 90–101

dies 13
diesel hammer 62–63
diffuser 59
Doblin, Jay 100
docks 158–165
dragline excavator *see*
 excavating machines,
 strip mining
drains 102–104
DRAINAGE 102–104
dredger 72–73
drop hammer 62
dug-out canoe 22
dumpers 69, 70–71
dynamometer 20–21

Egypt 22, 55, 78, 116,
 130–132
Eisenman, Peter 98
electric motor 14, 55, 60,
 61
electrostatics 14
English architecture 40,
 41–42 *see also* Jones,
 Wren, Stirling,
 TRANSPORTATION
 of houses
engine assembly 13, 16,
 19, 20, 21
ergonomics 90–93
Etruscans 10
excavating machines
 66–73
explosives 88–89
explosives (mining) 121
 see also
 DEMOLITION

Fiat 17
fibreglass 22–23, 26–27
Fitch, John 158
flier 53
FLOOD CONTROL
 105–108
Ford (cars) 16–17, 18–19
Ford, Henry 13
forgripe 23
foundations 47, 50
FRAME
 CONSTRUCTION
 14, 23, 24, 109–112
framing bucks 14
Frasch process 122
Fresnel, Augustin 118
Freyssinet, Eugene 44
Fulton, Robert 158

gabions 113–114

Galileo 29
galvanized steel 30
garbage *see* RUBBISH
geodesic dome 111
glass industry 123
glass reinforced plastic *see* fibreglass
Gothic architecture 9–10, 42–43, 54, 177
Greece, Greeks, architecture 10, 39, 40, 78, 137
grommets 14
Gropius, Walter 94
gypsum 78–80

Hamblin, Robert 117
harness (wiring) 12, 16
Heidegger, Martin 100
Hellespont 28
high-speed railways 36
Hitler, Adolph 94
Holland 113–115
Honnecourt, Villard de 49, 50
Hood, David 117
Hooke, Robert 29
houses (moving them) 168–169
Hutchinson, William 118
hydraulic excavator 66–67
hydraulic power 19, 37, 170
hydrograph 105

impeller *see* pumps
Incas 22
incinerator 149–150
India 113

Jencks, Charles 98
jib *see* crane
Jones, Inigo 43, 93, 96

kayak 22
keystone 9
Kitson, Arthur 116–117

laminated wood 11
LAND RECLAMATION 113–115
lapstrake construction 23
lathe, automatic 13
lathe, screw-cutting 12
Leibniz 95
lenses *see* LIGHTHOUSES
Leyland 19

LIGHTHOUSES and lightships 116–119
lintel *see* post and lintel
loader *see* excavating machines

McAdam, John 85, 139
Maudsley, Henry 12
Maya (building) 9
Mayflower II 24
megaliths *see* STONEHENGE
MINING and quarrying 120–129
moment (statics) 30–31
mortar 77

Napoleon 139

office furniture 92, 98–99
Oldsmobile 19
opencast mine *see* strip mine
outrigger 22

paint (on cars) 14, 16, 18
Palladio 29
Papin, Denis 158
parabolic curve 11
Payton, Joseph 44
penditive 40
Persia 28, 29
Peru 22
Petrie, Sir Flinders 132
photoelectric cell 146–147
piers 9, 10, 31, 36, 37, 46, 84 *see also* CATHEDRAL
piles, piledrivers 47, 62–63
Piretti, Giancarlo 98
plastics 46–47, 103
plinth 50
Plutarch 43
plywood 25, 45–46
post and lintel 9, 11, 39
post-tensioned concrete 89
Pontiac 18–19
pozzolana (cement) 28, 39, 77, 81
press room 13
primer (paint) 14, 18
priming *see* pumps
pumice 84
pump (bicycle) 55
pumps 56–61
putlocks, putlogs 53
pyradex construction 111–112

PYRAMIDS and stone circles 130–138
Pyroscaphe 158

quarrying 127–129
Queen Elizabeth II 162

radiation curing (paint) 14
Reformation 43
reinforced concrete *see* BRIDGES, BUILDING, concrete, DEMOLITION, ROADBUILDING
Renaissance 10, 29, 43, 44
reservoir 105–107
resistance welding *see* spot-welding
ROADBUILDING and street lighting 139–147
road safety 144–145
Roebling, John A. 30–37
Romans 9, 10, 28, 29, 77, 78, 84, 113, 116, 139, 176
Romanesque 40, 54, 148–150, 176
RUBBISH disposal 148–150
Rudyerd, John 118

Safdie, Moshe 95, 97
scraper *see* CONSTRUCTION MACHINERY, ROADBUILDING
sedimentation *see* sewage
SEWAGE and water supply 151–157
SHIPS and docks 158–165
slewing *see* cranes
Smeaton, John 78, 117
snore *see* pumps
sodium lights 146–147
Sommer, Robert 98
South Africa *see* MINING
spillway 105
spot-welding 14, 15, 16, 17
springing point (statics) 50, 54
squinch 40
statics 30, 31, 34, 48, 50, 54, 59
steamboats 158
steel 27
Stephenson, Robert 32, 36
Stirling, James 101
street lighting 145–147

stressed skin construction 109, 111
strip mining 124
STONEHENGE 133–137
Sullivan, Louis 44, 45
surface mining *see* strip mining
SURVEYING 166–167
Sweden 120, 148
Switzerland 28

Tacheometry 167
tantalum 59
tarmac 85
Taylor, F. W. 13
telescope 166
Telford, Thomas 29, 139
tellurometer 167
theodolite *see* SURVEYING
thermic lance 88
tides 108, 164–165
time study 13
titanium 59
TRANSPORTATION of buildings 168–169
Trésaguet, Pierre 139
TUNNELLING 170–175
typewriters 12, 90

underground trains 172, 174
unions, labour 13, 47, 101
Utzon, Joern 47

VAULT, the 10, 39, 176–179
Venice 12
Volkswagon 14
volute 59
voussoirs 9, 45, 179

water 151–157 *see also* DRAINAGE, FLOOD CONTROL, LAND RECLAMATION
water table 180–181
wells 180–181
Whitney, Eli 12
Winstanley, Henry 118
Wren, Christopher 43
wrought iron 32

Xerxes 28

ziggurats 39, 77

Picture Credits

American Motors Corporation 15T, 21T
Aspect 11, 45TR
Amy Roadstone Construction Ltd 86T
Architectural Association 96T&B, 97T&B, 101
Alphabet and Image 176R
Aerostyle/Paul Brierly 55B
Avoncraft Museum 42/3B
Anthony Dawson 73, 172L&R
A M & S Ltd Europe 123T
Actualit 126/7
Aerofilms 134
BSP Ltd 62BR
John Bishop 65T
Blakeborough 154
Blue Circle Cement 81B
Ron Boardman 85L
F E Beaumont 89BR
Bristol Corporation: J Porter 104T&BL, BR
Bristol Composite Materials 26T
BASF 48T, 82C
BAC 12T, 109
British Aluminium Ltd/Paul Brierly 142/3B, 143
British Steel Corporation 34/5, 46T, 110TL, 112BL&BR
Ben Rose/Scientific American 48B
British Museum 52BL
Broomwade Ltd 56L
Paul Brierly 56B
Derek Beckett 45TL
Colorific! 72, 122L&R
Corporation of London 145R
Clarke Chapman Ltd 149
CZ Scientific Instruments Ltd 166
Council of Industrial Design 90B, 91
Crown Copyright 108B, 144L&TR
Peter Clayton 131L
Richard Costain Ltd 141TL, 173T, 175BL
Compactors Engineering Ltd 142/3T
Cumbernauld Development Corporation 144BR
Douglas Dickens 180
Daily Telegraph 90T&C
Dave Hoskins 135T
C M Dixon 135B, 177
Ray Dean 120
Mary Evans Picture Library 32L&R, 124T&B, 170, 175TL
Esso 74L&R, 75TL&TR, 157T
Freeman Fox Ltd 33C

Peter Fitzjohn 51
Robert Estall 57, 133
Fotolink 67, 68TL
Roy Fooks 162R
Fairey Marine 25
Ford 15B, 16, 16/7, 18B, 18/9, 20, 20T, 94
Fiat 17B
John Freeman & Co 117T
Eugene Fleury 136
Fox Photos Ltd 108T
F H Lloyd & Co Ltd 71T
Susan Griggs/Adam Woolfit 105T&B
General Motors 18T, 17T, 21B
Michael Holford Library 35B, 36/7B, 52TL&TR, 77L, 130
Sonia Halliday 41T, 42TL, 176L
Highways and Road Construction 141BR, 142
Robert Harding Associates 131R, 131, 138B, 165L
John Hillelson/Bruno Barbey 43T
Howard Rotavators 103B
IBM 12B
International Harvester 70T, 70/1
International Labour Office 22/3
Ketton Rutland Cement Co 80
P Knowles 110TR
Keystone 65B, 168/9B
Leyland 19T, 19B, 59R
Laings Construction 89BL
London Brick Company 75BL, BR
William McQuitty 26BR, 35T, 69B, 83B, 102BL
Marshall-Fowler 68B
McAlpines 171, 173B
Nash Dredging & Reclamation Ltd 114T&B, 114/5T, 115TR&B
National Motor Museum 139L
Eric North 27B
Popperfoto 9, 24, 38, 108C
Picturepoint 27T, 33T, 45BR, 47B, 85TR, 88L, 112T, 113R, 116L, 165R, 175
Photo Library of Australia 31BL, 47T, 110BR, 121B
Photri 66B, 81T
Pynford Ltd 169T
S L D Oldring 140T
Ordnance Survey 167
Peter Reynolds 39
Ready-mixed Concrete Ltd 84R
Ruberoid Ltd 86B
Ronan Picture Library 87B
Radio Times Hulton Picture Library 102TR

G A Robinson 59L
Redland Brick Co 76
H Roger-Viollet 138T
Shell 145L
Kim Sayer 147
Staatsbibliotek 169B
Spectrum 128L, 129, 159, 161
Southampton University/Paul Brierly 82R
Slide Centre 29T
Scala 40, 42TL, 176L
Sheppard Building Group 52BR
Mike St Maur Shiel 64, 89T
Travenol Laboratories Ltd 60T
Trinity Lighthouse Service 118B
Thomas Green & Sons 141TR
Thompson Santer 146
Thorn Lighting 147, 146/7
Thames Water Authority 151L&R, 152TL&BL, 152/3, 153T&B, 155L&R, 157B
Vauxhall Motors Ltd 13, 14-15, 20B
Westerly Marine/Esso 26BL
Wolf, Laing & Christie 18C
Thos Ward Ltd 77R
John Watney 61, 117B&T
Vales Plant Register Ltd 62L
Zefa/Pictor 14, 68TR, 139R
Zefa 10, 28, 29C, 30T, 31T, 33B, 37, 42TR, 66T, 88R, 121T, 126L, 127T&B

Artwork Acknowledgements

Allard Graphics 83T, 113BL
John Bishop 65T
Diagram Visual Information Ltd 92, 102B, 106B, 106/7B
Eugene Fleury 136
Jackson Day Design 123B, 179
Frank Kinneard 58
Tom McArthur 55T
Osborne/Marks 34, 36/7T, 63, 78/9, 119, 125, 132, 140/1B, 150, 156, 160, 163, 181
Saxon Artists 11B, 23, 50/1
Thompson Santer 146, 174